GOLD PROSPECTORS

HANDBOOK

BY JACK BLACK

GEM GUIDES BOOK COMPANY
315 CLOVERLEAF DR., SUITE F
BALDWIN PARK, CA 91706

Copyright © 1987 by Robert F. Ames

1st printing, September 1980
2nd printing, February 1983
3rd printing, January 1988
4th printing, June 1991
5th printing, June 1993
6th printing, March 1996
7th printing, April 2000
8th printing, March 2005
9th printing, May 2008

ISBN 0-935182-32-2
Library of Congress Catalog Number 77-92558

Printed in the United States of America

TABLE OF CONTENTS

INTRODUCTION

Over thirty years ago when I first became interested in panning, dry washing and hard rock mining on a small scale, the really serious amateur prospectors of America could easily have held their conventions around a campfire—and often did. Today, their campers, trailers and tents constitute a population explosion in the deserts and mountains of the major gold producing states. And, when those with only a casual interest who pan or dig only occasionally are added to the total, the amount of persons interested in this activity are truly one of the larger groups of outdoor enthusiasts in the country. This activity is not entirely confined to the western states either. There is considerable interest east of Colorado, especially in the areas around Georgia. Alaska and Canada, of course, are also prime locations.

What is behind this tremendous growth in the numbers of weekend prospectors? Many authorities credit the development of the lightweight, portable gold dredge. Others cite the recent advances in the price of gold. The first was the spark that ignited the explosion, the price increase merely fanned the flames. The true story reveals much more.

One of the major reasons there are so many weekend prospectors today stems from the long-time artifically low price of gold. When the mineral was set at thirty-five dollars per ounce, virtually all commercial miners were driven from their occupation for economical reasons. By the early 1950's nearly all commercial gold mines in the West were closed. And this is when the weekend prospector started arriving on the scene. Another factor is the seemingly new discovery (although many of us have known it for years) that there is still plenty of gold left in many western streams. Literally every time it rains, or when winter snows melt, more gold comes down to replenish that taken out previously. Despite all the weekenders taking home their "ounce" each season, there is still enough left for anyone with the patience and know-how to find it.

But even though these reasons are valid, I would like to point to the single reason I think most responsible for the continuing popularity of amateur prospecting. This is simply the thrill of finding your first show of colors, or your first nugget.

A poet once wrote that men searched not for the gold, but for the finding. Once you have run your fingers through black sand or gravel to turn up a small nugget, you will know this feeling, too. Despite its monentary value, and your first find may be worth very little cash, the nugget will represent a feeling of accomplishment that little else can come close to. Then you will know why so many persons are spending their spare time panning, dry washing or digging.

While the newcomer will find elementary advice easy to get, and the principles quite simple to master, there is a great gap in the literature of amateur prospecting. There are many short articles and small pamphlets that briefly introduce the subject of prospecting, but these stop just where they should begin. On the other end of the scale, there are excellent engineering books which describe in highly technical terms the principles of professional mining and prospecting.

This book is written to fill this void. To show the beginner how to get started and progress to the point where he can become an advanced, proficient amateur at prospecting and small mining operations. To teach him how to be one of those who can bring home the bacon—or gold, mercury, silver, or whatever else it is he seeks.

JACK BLACK
Canoga Park, California
1980

1

WHAT IS GOLD?

To many persons gold is just a word. It is the metal with which their jewelry is made, with which their teeth are filled, or it may be just an item in the paper telling how much per ounce it is selling for today.

But to the prospector it is far more than just another metal. For him it is a dream alive with memories, hopes and realizations. If you want to be successful in weekend prospecting, some of this "aura" of gold should first rub off on you. The most successful prospectors have always been those who were motivated more by the finding than the actual profit they could make.

To understand just why it is so much sought after and why a pastime such as amateur prospecting could evolve around the metal, it is necessary to study a few of its peculiarities first. Very simply, gold is an element and almost always is found as virgin metal (there are exceptions such as the tellurides, which are seldom the concern of the recreational miner). One of gold's most distinguishing features is its color, a pale to deep yellow, which stays nearly the same whether the mineral is scratched, melted or cooled.

When hammered, gold will become flat; other similar yellow minerals break or disintegrate. On the scale of hardness, gold is very soft, ranging from 2½ to 3 On the other hand, it is very heavy, with a specific gravity slightly more than nineteen. A cubic foot of pure gold weighs about twelve-hundred pounds avoirdupois; and a cubic inch about eleven-and-a-half ounces troy. At early 1980 prices a cubic inch of gold would be worth from $6,000 to $7,000.

Gold can be pounded so fine that 250,000 thicknesses are required to make an inch. In laboratories it has been hammered to a thickness of five-millionths of an inch. It can also be broken down so fine that 2,000 particles are required to make a quarter-cent of value (of course, it was a cent at $35 per ounce). Even particles this minute can be readily recognized in a gold pan and economically recovered. Gold is so ductile that an ounce of it can be drawn into a wire fifty miles long, or that same ounce will plate wire the thickness of a thread one thousand miles long.

Gold is, then, a most unusual metal. It is so easily worked that primitive man quickly recognized its value and began using gold as long as six thousand years ago. Of these sixty centuries, only the last one hundred and thirty years are of any real significance in the history of mining gold. In this period more than three-fourths of the gold now in existence was mined. Two-thirds of the world's gold supply has been mined since 1931. If it were possible to gather all the gold mined in the world since Columbus first discovered America, it could be cast into a cube with dimensions slightly more than fifty feet. The total weight would be about 73,000 tons.

No building would be necessary to house the gold for it has yet another property that has endeared it to men since ancient days. It is practically indestructible and will not tarnish. Just a few years ago a skin diving group brought up coins from a Spanish galleon which had sunk in the 1700's. These coins had been freshly minted in Mexico City before being shipped and had never been circulated. When cleaned they looked like they had just been minted the day before instead of almost three centuries earlier.

Finally, gold is money. It always has been, and despite what politicians and economists try to put over, it always will be. Yet, a million dollars is an insignificant lump of metal. At $600 an ounce, it would be a cube approximately 5.5 inches on each edge and would weigh about 240 pounds. Cast into two pieces, a million dollars could be easily stored at home or transported in a hurry in case of disaster. This simple fact is a positive proof why gold is currently so popular in politically unsettled countries of the world.

Another aspect of gold's personality that has made men desire it over the years is its constant value. For centuries the price of gold has steadily increased and has never gone down. Even though gold prices have fluctuated wildly in the past decade and will probably do so more in future years, it has reached substantial levels and has not retreated. There is no doubt that gold will soon find a level at which the price will stabilize and this general price level will be sustained for decades.

To see how stable gold has been in the world's economy one economist traced its rise in price over a more than 600 year period. Before the dollar was established in the late 1700's, he made his comparisons to the English pound. Here is the table he made brought up to date.

DATE	YEARS OF CONSTANT PRICE	PRICE PER OUNCE
1344	7	$ 5.58
1351	91	6.37
1442	22	7.18
1464	62	8.03
1526	78	11.02
1604	7	15.90
1611	52	17.36
1663	36	19.82
1669	18	20.36
1717	217	20.67
1934	38	35.00
1972		38.00
1980	Summer	650.00

This, then, is what gold is, and partly explains the tremendous interest by amateurs in prospecting for, and mining of, gold.

Gold is man's most sought after commodity—his most prized possession. With a little luck and the things you learn in this book, you may be able to find a little, or perhaps even a lot.

But, even if you find only a trace, you will have accomplished something few men have done in modern times. You will have personally discovered something of instant value from nature's vast storehouse of treasure. To others it seemed like a simple rock, or river sand. However, when you concentrated the gold, it instantly became cash even better than that in your wallet.

This is the thrill of prospecting. Finding gold is far more rewarding than later converting it to cash to spend—unless you find a hundred ounces or more, and then spending is very rewarding.

2

IS THERE ANY GOLD LEFT?

The answer to this question is a resounding, yes! There is probably more gold remaining than has ever been taken out. This is especially true of California. The reasons for this are many. One has already been discussed—the economics which forced professional miners to abandon their workings. But a more important reason—replacement of placer deposits—should be most encouraging to all potential prospectors.

To the uninformed it often comes as a surprise that many stream placers constantly replace themselves. The way this occurs will be examined in depth in later chapters but a brief description here will prove beyond all doubt that there is still plenty of gold left to be recovered.

California's Mother Lode country is a good example. Eons ago there were many rivers and streams which were concentrating gold before man appeared on the scene. When the Sierra mountain chain was formed by thrusting upward, these rivers were left high and dry. Other streams and rivers formed, cutting directly across the stream beds of the former rivers.

The beds of these ancient rivers contained much gold that had been accumulated over the centuries and the new rivers started transporting it to lower elevations where it concentrated at bars and near bedrock. Naturally over a period of centuries a tremendous quantity of gold could be accumulated this way and it was this accumulation that the forty-niner found.

Within a few years these placers were worked out and the miner who wanted to make a living had to turn elsewhere to find gold. This is when the lodes and

Tertiary river channels were discovered. Due to economics and legal problems which closed down the hydraulic mines in the 1880's, the tremendous Tertiary river channels with their reserve of low grade gold deposits are no longer worked by anyone but Mother Nature herself. Each time it rains, or each time snows melt, gold is washed to a lower level where it is concentrated in streams and rivers.

This constant replenishing of gold placer fields goes on in many locations besides the Sierra's of California. But the implications of the process are plain— for centuries to come there will always be some gold in placer locations. There are bound to be some areas that produce only small returns as they are constantly worked by amateurs. The headwaters of Piru Creek in Southern California is a good example of this. On the other hand, human nature being what it is, there will still be some areas that will be overlooked for many years by everybody until some weekend prospector decides to take a chance. Less than ten years ago, three weekend prospectors took such a chance and recovered over twelve troy pounds each in a very short period of time. There will always be both relatively small placers and those who produce relatively large bonanzas as long as the laws of nature work.

Just how much gold is there left? Well, this might be anybody's guess but a look at past performance will give a clue. The gold fields of California have been accurately recorded for years. The production records shed some light on the question. (Note: The mining departments of most other states have reported similar facts for this period so the argument will hold for most states.)

The last big decade of placer mining activity in California was from 1931 to 1940 when thousands of unemployed persons flocked to the gold bearing streams. The total amount of placer gold recovered in California streams during the period was $134,971,000.

This is an impressive figure even today, but it proves only that a considerable amount of gold was recovered during this period. Fortunately, during the years from 1931 to 1940 the California Division of Mines was very sophisticated and segregated the types of placer mining. The lion's share of placer gold was produced by commercial operations. Small scale placer miners still were able to get about 8.5%, or nearly $11,400,000 worth of gold.

Adjusting for the price of gold at that time, this means that small time placer miners recovered about 330,000 troy ounces or about 27,500 troy pounds of gold working by hand methods. Considering the drop of placering activity during the 1940's and 1950's, which allowed many of the streams to replenish themselves with gold, it is not inconceivable that there is still from 2,500 to 3,000 pounds troy of gold to be recovered each year from California streams alone. If gold reaches the predicted $1,000 per ounce price, this would mean $300,000,000 per year in recovered gold.

Before this information starts a new stampede to the gold country let's examine a few other figures compiled by myself using some other statistics given by Charles V. Averill about this period. He estimated that there were as many as 6,000 small time operators in the field at this time. Taking a three month season, this breaks down to an average of slightly more than five ounces per man or about $175 per season. This is slightly more than the figure given by Averill. In his estimate he placed the average recovery at $3.50 per week.

The amateur prospector who wants to get going in panning, sluicing or dredging should not be too discouraged by these bleak facts presented in the previous paragraph. Averill himself pointed out that the peak of 6,000 prospectors brought a horde of inexperienced, inefficient people who knew little or nothing about placer mining. Most of these men became discouraged after a few months and never returned to the fields. About 1936, his report states, only those who knew their business remained. Now that they were free to work the sand bars, the professionals started to get bigger and bigger shares of placer gold.

For California, at least, there is as much as 3,000 troy pounds of gold to be shared each year, so there is plenty of gold left. And, I believe, as did Charles Averill, an experienced mining man, that the biggest share will go to those who know where to look and the most efficient method of getting it out.

3

ELEMENTARY GEOLOGY

Before any study of the basics of geology as it applies to the prospector it is necessary to present a brief understanding of minerals and how they are formed. Miners and geologists are often at odds when they use the term mineral—for there are many definitions of the word used in various professions. Here, minerals will be described as elements, or compounds of elements, which are formed naturally within the earth and later brought to the surface by geological forces. By this definition metal will be considered a mineral although many scientists and miners do not agree with this terminology. Rocks, on the other hand will not be described as minerals but as a combination of various minerals, or elements.

It is also good to have at least a speaking acquaintance with the science of chemistry. Actually, it is enough to know that all matter is made up of elements. Scientifically speaking, an element is a substance which cannot be broken down further into simpler substances by a chemical process. This excludes atom smashing which is not considered chemical.

Some elements such as aluminum, gold, silver, iron, etc., are considered a mineral, but most minerals are compounds of two or more elements.

By weight, 98% of the crust of the earth is made up of only eight of the elements: Oxygen, about 47%; silicon, about 27%; aluminum, about 8%; iron about 5%; calcium, about 4%; and with percentages ranging over 2% are sodium, potassium and magnesium. The balance of the crust is composed of the other elements.

Although the mineral forming process is still going on inside the depths of the earth, virtually all of the minerals found in recent history were created millions and sometimes billions of years ago. Some persons are confused on this mainly because many minerals have only been exposed in the last million years.

Minerals were formed deep in the earth in materials called magma—a portion of the under-surface earth that is still molten. These areas are found in the crust and mantle of the earth and many are still active as shown in the phenomena of volcanos.

While not the most scientific method available, the formation and action of minerals after they are formed can be compared roughly to the action of minerals in the gold pan, the sluice or the stream.

In the beginning the magma is liquid and heated to fantastic temperatures under extreme pressure. As the magma cools, elements are joining together with themselves or with other elements to form compounds. As these new mineral forms begin to cool they take the shape of crystals. These crystals vary in size from microscopic to tremendously huge. Once the minerals have been created, the lighter ones begin "floating" in the magma to the top or side nearest the surface of the earth. Heavier minerals begin sinking to the bottom while those minerals whose weight is in between are suspended. Gradually, sometimes the process consists of millions of years, this process is completed. Sometimes afterward, often while the material is still in a molten stage, the whole of it is thrust upward through geological processes.

The geological processes which bring minerals to the surface will be discussed later in this book, along with the geological forces which act on them once they have been exposed.

We can now begin examining a scientific procedure to identify and classify major divisions of the huge family of minerals. Although the average reader will be most interested in gold, it is important to be able to know how to identify other minerals in the field. Many quests for gold have turned up some other valuable mineral, but only because the prospector knew how to identify the substance he found.

Crystals are one of the most important identifying characteristics of minerals. Well defined crystal shapes are formed when crystalline minerals are created without interference from other pressures. All crystals have been assigned to six *systems.* These systems are identified by their axes; i.e. a vertical axis runs from top to bottom and a horizontal axis runs from side to side.

Isometric or Cubic System. These crystals have three axes of equal length at right angles to one another. It forms a near perfect cube.

Tetragonal System. Like the Isometric System, Tetragonal crystals have three axes at right angles. The horizontal axes are of equal length but the vertical axis is longer or shorter than the horizontal.

Hexagonal System. This six-sided crystal has three horizontal axes of equal length and a vertical axis running at right angles. The vertical axis can be either longer or shorter than the horizontal.

Orthorhombic System. These crystals have three axes all at right angles but each has a different length.

Monoclinic System. In this system there are three axes of unequal length. Two of them intersect at right angles and the third axis has an angle that is oblique to one of the others.

Triclinic System. Triclinic crystals have three axes of unequal length and all intersect at oblique angles.

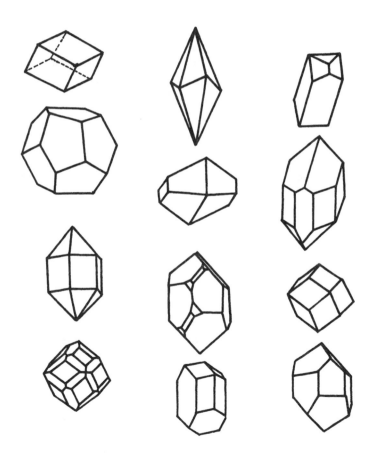

Some common forms of crystals as found in nature.

Besides the systems with which crystals are identified, they also form in different, well defined shapes called *habits*. Three of the most common are: column, cubic and tabular. This is not as positive an identifying feature as it might seem. Temperature, pressure, composition of the solutions from which the minerals crystallize, variation in mineral composition and impurities in a mineral may change its habit. For instance, fluorite crystals, which are a member of the isometric system, may be in the cubic habit if formed at lower temperatures but often appear in the octahedral form if formed at much higher temperatures.

Crystals observed by the eye, or under a magnifying glass, often positively identify a mineral, but more often a mineral will defy identification this way because, due to changes caused by fracturing or erosion, the smallness of the crystal, or chemical action on a metallic crystal.

Over the years geologists and prospectors have developed a basic system of testing the physical properties of minerals in the field so that they can be identified with only elementary scientific equipment through a process of elimination. These properties, and simple tests for them are listed below.

Hardness. One of the most important and easiest methods of putting any mineral in a general class is to find out approximately how hard it is. This is determined by what materials it will scratch and what materials will scratch it. Hardness is rated by Moh's scale. Devised by a German mineralogist, this is not a mathematical progression of relative hardness but merely assigning numerical progression to ten different minerals arranged in varying degrees of hardness. It is as follows:

Moh's Number	Mineral	Moh's Number	Mineral
1	Talc (softest)	6	Feldspar
2	Gypsum	7	Quartz
3	Calcite	8	Topaz
4	Fluorite	9	Corundum
5	Apatite	10	Diamond (hardest)

Kits are available in most rock shops which will test hardness but most prospectors buy the various minerals. They are not really very expensive if gem quality minerals are not purchased. With the advent of synthetic gems, small quantities of corundum (ruby) can easily be acquired. Diamond can usually be obtained in the shape of a small rough diamond mounted in a metal bar which is used to dress grinding wheels. If a prospector is in the field without a hardness kit many common items will give a rough approximation of the hardness of most minerals. All the following hardness ratios are approximations only: Fingernail, 2½; Copper penny, 3; Glass, 5-5½; Pocket knife blade, 5½-6; and a steel file, 6½-7.

To make a test for hardness most prospectors start with softer tests first. If it is obvious that it is a soft mineral, they try scratching it with their fingernail. If this does not produce a scratch, they generally jump immediatetely to a pocket knife or a piece of quartz. If the quartz produces a scratch, they proceed to a topaz test. If the topaz does not scratch the mineral being tested and the quartz does, it indicates a hardness between 7 and 8. Usually this is as far as it is necessary to carry the test. If the mineral tests 8 or higher, most prospectors take it it in for a professional assay for it is bound to be something of value.

Unfortunately, proving that a mineral you have found is between 7 and 8 in hardness proves little, most of the minerals which you find and test will fall into this range. But once you have determined hardness, there are several other tests which will help identify a find in the field.

Color. When a mineral is found, the first thing that is usually noticed about it is the color. However, since many minerals have the same, or nearly the same, color the hardness test is usually performed first. Then elimination of possibilities by color is begun.

It is important to consider things which might change the color before checking your reference tables. Natural and artificial light, wetness, fresh surfaces and weathered surfaces can all cause slight variations in color.

Streak. Mineralogists long ago discovered that when a mineral is pulverized to a powder its color often changes markedly. A good example is black hematite which turns to a reddish brown when powdered. To standardize this form of identification, minerals are drawn across a piece of unglazed tile called a streak plate. This action can be best described as drawing a charcoal pencil across a rough piece of drawing paper changing much of the charcoal into a line of black powder. Instead of a pencil a piece of mineral is drawn across the unglazed tile and the resulting powder line is called the streak. The color of the streak is checked against the tables for various minerals and a further elimination of possibilities is made.

If no streak is produced during the test the material is probably harder than 7 as this is generally the hardness of the streak plate. While almost any piece of unglazed tile can be used for the streak test, it is better to purchase a standard tile from a rock or prospecting shop as these are of the right color. Since the powder is normally viewed on the streak plate, any background color of the tile can contribute to the final color of the streak.

Luster. When minerals are viewed in reflected light their appearance is said to reveal luster. Judging some classes of luster can be quite easy—such as the difference between metallic and dull. Other divisions of the luster test must be acquired through experience as some of the differences are slight.

The best way to learn how to judge luster is to obtain several known mineral specimens of all the types of luster classifications and keep them as labeled samples. It is important that the lighting be the same or nearly the same at all times. Bright or artificial light can greatly change luster. If the specimen is viewed on a cloudy day, or on a day with the sun shining while the specimen is in the shade, the luster will appear different.

As a big generalization, luster is divided into two general classifications—metallic and nonmetallic. The latter class has so many subdivisions that it is really the identifying part of this test. Here are a few of the classifications each followed by a common example of the luster class.

Metallic. Shiny like metals in their natural form—silver, gold.

Submetallic. A classification between metallic and nonmetallic.

All the following are divisions of *Nonmetallic*

Adamantine. Brilliant glossy luster—diamond.

Dull. Very dull appearance—chalk or clay.

Greasy. Like an oily surface—nepheline.

Pearly. Like silk or rayon—abestos.

Resinous. Like resin—sphalerite.

Silk. Like silk or rayon—abestos.

Viterous. Like glass—quartz, topaz.

Specific Gravity. This well known physical property will eliminate many possibilities and is often performed right after the hardness test. Scientifically the test is performed by weighing the weight of a mineral specimen with the weight of an equal volume of water. Due to the many unusual shapes of minerals, determining the volume of a specimen is virtually impossible.

However, the specimen can be weighed in air on a spring scale and lowered into water taking a second weight reading. The specific gravity is determined by dividing the weight in the air by the loss of weight in water. For instance: if the specimen weighed 20 ounces in the air and 16 ounces in water, the specific gravity would be 5.

This test is only an approximation as the accuracy of the scale and the pureness of the water can affect the outcome. If performed in the field be certain that the water comes from a fresh water source and not a saline lake as salt can reduce the weight in the water-weighing enough to be off in computation.

Clevage and fracture are two more methods of identifying minerals. Unless they are metals, most minerals will break if they are struck a sharp blow. If they break irregularly, the mineral is said to fracture. If it breaks along regular lines or patterns, it exhibits cleavage. Some of these peculiarities are very valuable in identifying minerals—but, because so many minerals fall into the same classification, some of them are next to worthless.

Cleavage. If a mineral is struck a sharp blow with a hammer it may shatter into numerous small pieces. If these pieces have lines corresponding to the mineral's original crystal structure much can be learned. For instance, galena crystals have three axes of equal length and at right angles. When shattered galena breaks up into small cubes.

Fracture. When struck with a hammer many minerals shatter in a distinctive way even if they do not show cleavage. When this happens the mineral is said to fracture and the broken surfaces are sometimes of great value in identification.

There are many types of fractures but only a few are of interest to prospectors. Here are the most important with a common mineral listed to illustrate each:

Conchoidal. Breaks like the smooth surface of a sea shell. Similar to those often found in thick pieces of broken glass—obsidian and most quartz.

Splintery or Fibrous. Fibers or splinters are found on the surface—pectolite.

Hackly. Surface marked by rough, jagged torn edges—copper, gold, silver and other metals.

Uneven. Fractures rough and irregular. This characteristic is common to so many minerals that it is almost of no use in identification.

Another identifying characteristic of many minerals is *Tenacity,* the resistance it offers to bending, breaking, crushing or tearing. With examples, the most important classifications of Tenacity are:

Brittle. Brittle minerals can be broken or powdered with ease—galena, sulfur.

Elastic. The mineral returns to its original form after being bent.

Flexible. Bends, but does not go back to its original shape—talc.

Sectile. This type of mineral can be cut with a knife in the manner of whittling wood—gypsum, talc.

Malleable. When pounded with a hammer the mineral can be reduced to thin sheets—gold, silver, copper.

Ductile. Some malleable minerals can be drawn through a small hole producing wire—gold, silver, copper.

As can be noted from the previous listings, many minerals exhibit more than one characteristic of tenacity. This is a great help in identifying some of the more difficult to identify minerals.

In addition to the more common tests already discussed some mineral identification guides add a few of their own. Some of the more widely used by prospectors are:

Play of colors. Many minerals will reveal different colors or grades of colors when they are rotated in a steady light source—labrodorite.

Asterism. Highly prized for gems, these minerals exhibit a star when the light strikes them at the correct angle. The phenomena can be viewed at most rock shops by asking to see a Linde Star, which is a synthetic gem stone—star rubies and sapphires.

Luminescence. Entire books have been written on this subject and it is best left as a specialized type of prospecting. Some minerals glow when exposed to ultraviolet rays (Fluorescent); while others continue to glow in the dark after a light has been removed (Phosphorescence).

Magnetism. Many minerals are composed largely of iron and can be attracted by a magnet. In some cases (lodestone) the mineral itself will act as a magnet. Magnetite, often a nuisance in gold placering is often removed from concentrated sands with a magnet.

Transparency. This refers to the ability of light to pass through a mineral. *Opaque,* means no light. *Translucent,* some light passes through but an object cannot be seen through the mineral. *Transparent*, light passes through and objects can be viewed through the mineral.

All of the previously mentioned tests are the basic methods of identifying minerals but you should be aware that there a few exceptions to the crystal families of minerals. Most minerals solidify as crystals but others lack the ability to crystallize and are found as a hardened gel. Some writers refer to them as mineraloids but the most common term is amorphous. The best known specimen would be opal.

This chapter will not make you an expert on mineral identification. But it should give enough information to identify most common minerals you find on your prospecting trips. A later chapter, Other Valuable Minerals, has several of the more common valuable minerals listed together with their places in the various classifications of identifications.

4

ELEMENTARY MINERALOGY

Like learning about the science of mineralogy, the weekend prospector does not need an extensive knowledge of geology. Therefore this chapter will be greatly condensed and present definitions and generalizations only. Most amateur prospectors do their looking in the same general area and soon become specialists in the geology of that area. Having an understanding of some of the more basic facts will help grasp much of the published geological information about the area you are interested in.

One of the most important things to learn about geology is the way the geologist thinks of time. When the geologist says it took a hundred million years for some mineral to form, he thinks of it as a relatively short time. To the average person one year is a very long time.

Think of the five billion years the earth has been in existence as having occurred in one of our modern years that was 365 days long. Each minute would be about 10,000 years long and a second would take 158.55 years to pass. The Recent Epoch, about 11,000 years long, would pass in one minute and ten seconds. The entire history of man since the birth of Christ would be over in less than a quarter of a minute; and World War II would take only four-hundredths of a second, less than the time it takes to blink your eye. If you can think about geological processes in this manner, you will find it easier to understand the way minerals and lode deposits were formed.

GEOLOGICAL TIMES

PRECAMBRIAN ERA. The Archeozoic and Proterozoic Periods comprise one Era called the Precambrian. This makes up the oldest of all geological activity and is considered to have occurred from 600 million years back to at least two billion years. The Era may have lasted over four billion years—almost 90 percent of the earth's known geological history. As science progresses the age of the earth is being considered much older and modern scientists believe that our planet may well be over five billion years old.

The Precambrian Era was marked with colossal and even fantastic earth changing events. Much land was at times under water, other areas were fired by volcanic activity and mountains were uplifted. Despite the many difficulties in forming scientifically proven theories, it is evident that many of the minerals and metallic ores were formed at this time. Rocks of the Proterozic Period have given up large deposits of gold, silver, iron, copper, nickel and cobalt, both in the United States and Canada.

PALEOZOIC ERA. This Era encompasses that period of time that began about 600 million years ago and ended about 200 million years ago. At the beginning the first marine life, trilobites and brachiopods, made its appearance. The end of the Era roughly corresponds with the appearance of the dinosaurs.

The Paleozoic Era is divided into six Periods which in ascending order are: Cambrian, Ordovician, Silurian, Devonian, Carboniferous and Permian. The Carboniferous period is divided into two Epochs: the oldest, the Mississippian; and the latest, the Pennsylvanian.

All during the Paleozoic Era there was considerable geological activity. North America was drained of many of the seas that covered it. Following times of uplift, when mountains and hills were formed, the land was once again submerged and at times much of the world was under water or swampy. It was during this Era that many of the world's oldest coal deposits and fossil fuels were formed.

MESOZOIC ERA. This Era lasted until approximately 65 million years ago and is divided into three Periods, the oldest being the Triassic followed by the Jurassic and Cretaceous Periods. The Era began with the appearance of the dinosaur and ended with the development of early form mammals.

During the Mesozoic Era animal life multiplied at a fantastic rate. During this Era major portions of the North American continent were submerged for the last time. The Sierra Nevada mountains which begin in California and Nevada and extend into British Columbia were formed. It was also during this time that igneous intrusions formed some of the world's richest gold producing veins. During the Cretaceous Period the Rocky Mountains were formed. As plant and animal life were buried in huge numbers many more of our fossil fuels were created.

CENOZOIC ERA. This Era lasted for approximately 60 million years and is of considerable interest to gold prospectors because it is during this Era's two Epochs (the Tertiary and the most recent, Quaternary) that the rivers were formed and reformed to leave mammoth deposits of placer gold which have not yet been anywhere nearly exhausted.

The Tertiary Period is divided into five Epochs. Starting with the oldest, they are: Paleocene, Eocene, Oligocene, Miocene and Pliocene. This point in time saw the emergence of the first placental mammals and ended with the appearance of large mammals.

The Quaternary Period began with the Pleistocene Epoch and brings the earth's geological history up to date with the Recent Epoch. During the Pleistocene Epoch it is believed that man first appeared on earth although recent discoveries may reveal man to be even older. The Recent Epoch covers the last 11,000 years of man's modern history from his earliest beginnings to the events of yesterday.

The Cenozoic Era is marked with abundant animal and plant life mostly in the forms that we know them today. Lasting about 60- to 65-million years, it was also another time of great geological activity. Many of the great oil fields in California, Texas and Louisiana were discovered in Tertiary rocks. Also volcanic activity and great crustal shifts formed the Coast Range of California and the Cascade mountains of Oregon and Washington. Much of the coal mined in the western United States is of Tertiary age and the copper, gold and silver deposits of Bolivia, Peru, Mexico and the Rocky Mountains of the United States were formed.

The later portion of the Pliocene Epoch saw the beginning of the glaciers, giant ice sheets which covered much of North America, Europe and Siberia. The beginning of modern geological history coincides approximately with the last great Ice Age, about 11,000 years ago.

ROCKS AND HOW THEY ARE FORMED

IGNEOUS ROCKS. While it is presently believed that most of the mantle just beneath the crust of the earth is now solidified, there are still many areas of molten rock. When this rock cools and solidifies it forms minerals and rocks. The molten area is called magma. When a volcano erupts it is spilling magma onto the surface of the earth; when the lava cools it is an igneous rock. This process of delivering magma to the surface is called extrusive.

But far more magma has been brought to, or nearly to, the surface of the earth by being forced into other rocks where it then solidified. This action is called intrusive or plutonic rocks. As intrusive rocks cooled and solidified they produced most of the minerals and metals that prospectors search for today.

Volcanic rocks are less important to the prospector than the intrusive igneous rocks as it is these latter which have formed most of the commercial mining deposits that have been discovered.

Intrusive rocks are virtually injected into open or weak places in the earth's crust by pressures from below. The lower material cools slower and produces a coarser texture, the magma near the surface cools quickly and produces fine grained rocks. Sometimes the intrusive rock will cover portions of the earth's undersurface for thousands of square miles and becomes the country rock surrounding the lodes that are mined. Some common forms of these great deposits are gabbro, granite, periodotite and syenite. The most often found extrusive rock is basalt, also a country rock in many mines.

Batholiths. Largest of the igneous intrusions, they are irregularly shaped. Batholiths cover great areas and some extending for thousands of square miles have been discovered in Idaho, the Rocky Mountains and beneath the Sierra Nevada mountains of California.

Dikes. When magma is injected into sedimentary rocks and forms an extension cutting at an angle across these rocks, it is called a dike. When it goes straight up it is usually called a volcanic neck and seldom exceeds a mile in diameter. Dikes, on the other hand, can range from a few feet thick to several hundred feet and often extend for miles.

Laccoliths. Another massive form of intrusive magma which can cover thousands of miles, laccoliths resemble a dome with a straight, or slightly curved bottom.

Pegmatites. These deposits are found at or near other larger igneous rock deposits. They are often small, some only a few inches, to a few feet, in length and come in many shapes, lenslike being the most common. Smaller pegmatites often contain valuable gem crystals such as garnet, topaz, beryl, emerald, tourmaline and sapphire. Larger ones frequently contain lithium, fluorine, niobium, tantalum, uranium and some rare earths.

SEDIMENTARY ROCKS. Any rock, igneous, metamorphic or sedimentary when exposed to the earth's surface is subject to attacks of chemical and mechanical agents of erosion. When eroded to small size it can be transported by wind, water or ice and deposited in layers called strata. These layers are often hardened into a new stone by a process called lithification and the result is called sedimentary rock. Sedimentary rocks are classified in two forms: clastic or chemical.

Clastic Sedimentary rocks are those made up of rock fragments which have been derived from other decomposed or disintegrated rocks. Some of the more common sedimentary rocks are: shale, sandstone and conglomerate.

Chemical sedimentary rocks are those which have been precipitated from material dissolved in water. Inorganic chemical sediments are those dissolved in water and deposited by evaporation. Examples of this type would include rock salt from sea water and some types of limestone, dolomite, etc. Organic sedimentary rocks are those deposited by the remains of plants or animals which were not dissolved in water. Examples of this type would be coal, diatomaceous earth and some types of limestone.

Sedimentary rocks have also given us a scientific classification of the sizes of rocks themselves. These classifications are important to prospectors as much mining literature use the terms to describe mining operations.

ROUNDED, SUBROUNDED AND SUBANGULAR ROCKS

FRAGMENT	SIZE	AGGREGATE
Boulder	Over 256 mm	Boulder, gravel or conglomerate
Cobble	64-256 mm	Cobble, gravel, cobble conglomerate
Pebble	4-64mm	Pebble gravel, pebble conglomerate
Granule	2-4 mm	Granule sand
Sand	1/16 mm-2 mm	Sand, sandstone
Clay	Under 1/256 mm	Clay, shale

METAMORPHIC ROCKS. The third major classification of rocks sounds a little like the sedimentary classification but the major difference is that metamorphic rocks were not deposited nor were they broken down by erosion. This classification includes those rocks which were originally igneous or sedimentary and were changed radically when subjected to tremendous pressures and temperatures. There are two basic types of metamorphism.

Contact metamorphism occurs when an area is intruded by a massive quantity of magma. As it cools the magma produces a zone which according to geological terminology is baked. This takes place while the magma cools and brings about great chemical changes, and often re-crystallization, in the rock which has been intruded. In the event the rock intruded were limestone, it would be metamorphised into marble. This baked zone can extend for only a few inches or several miles depending upon the conditions when the magma intruded and first began to cool.

Dynamic or kinetic metamorphism is the second classification of metamorphic rocks and usually occurs when layers of rock are uplifted to form mountain ranges. During this process great changes in the underlying rocks are brought about due to the changes in weight and the heat and pressures produced.

Some of the changes in sedimentary rocks would include: sandstone to quartzite; shale to slate; or, bituminous coal to anthracite coal to graphite. Granitic igneous rocks become gneiss and the more compact igneous rocks become schist.

5

GEOLOGICAL AGENTS

After understanding how minerals and rocks are formed the most single important group of facts that a miner should absorb are those concerning the crustal movements of the earth. These are gradation, vulcanism and tectonism.

Gradation refers to the breaking down of surface rocks by air, ice and water. It is broken down into two important subdivisions: degradation; a wearing down, and aggradation, a building up by deposition. Usually these two effects are discussed intermingled under the terminology, weathering, as there is no really satisfactory method of separating them. Most of these forces are still working and the process has been continual for billions of years.

Physical weathering covers the process when rocks are reduced to smaller fragments without undergoing a change in chemical composition.

One of the more important causes of physical weathering is the formation of ice in cracks. Geological writers have commonly referred to glaciers and their actions as "ice." In the case of annual ice formations they refer to it as "frost." A process of expansion, frost action is broken down into two subdivisions; frost wedging, where ice forms in pockets or depressions and exerts a sidewise force; and frost heaving which describes the process when ice is formed in cavities beneath the earth and exerts an upward pressure. Wedging can be observed in fissures, cracks and potholes. One of heaving's most notable effects is the uplifting of road surfaces during winter. Due to the annual expansion and contraction of either phenomena, rock is slowly disintegrated over a period of many years.

Even if the rock area does not have a porous section to accept water for the frost action, alternate heating and cooling can in itself help to cause disintegration of rocks. The alternate expansion and contraction of a rock surface, while probably not a major disintegrating force of great consequence will produce small cracks and fissures where frost can invade and greatly speed the process.

Water, while mostly an agent of deposition, also is one of the major contributors to gradation. Everyone is familiar with the effect of water cutting a small pothole in solid rock, the result of thousands of years of dripping. So while water is mainly an agent of deposition, it causes erosion of the rock at the same time.

Plants and animals are also agents of erosion and cause the disintegration of rocks in many ways. Trees, bushes and smaller plants often grow in small crevices in rocks where their roots cause lateral forces spreading the crevices wider allowing larger frost pockets to form and occasionally pushing the rock surface apart exposing more surface to erosional agents. Burrowing agents such as mammals, rodents, ants and man himself bring rock particles to the surface exposing them to the forces of disintegration. Many miners have made big finds by observing the burrows of rodents and the materials they expose. And, in several states, the hydraulicking of the late 1800's exposed many Tertiary deposits which have been eroding gold particles for well over a hundred years.

The previously mentioned effects of physical weathering only produce smaller fragments of the rock and exposes them to other forces which may then transport them or change them.

Chemical weathering causes a breakdown of the structure of the rock and frequently creates a new mineral. This phenomena is also called decomposition and the processes most important to miners are *oxidation, hydration, carbonation* and *solution.*

Almost anyone can understand oxidation. It is the process whereby oxygen combines with minerals to form a new compound. It is easily illustrated by observing a piece of iron which has rusted after being exposed to the weather.

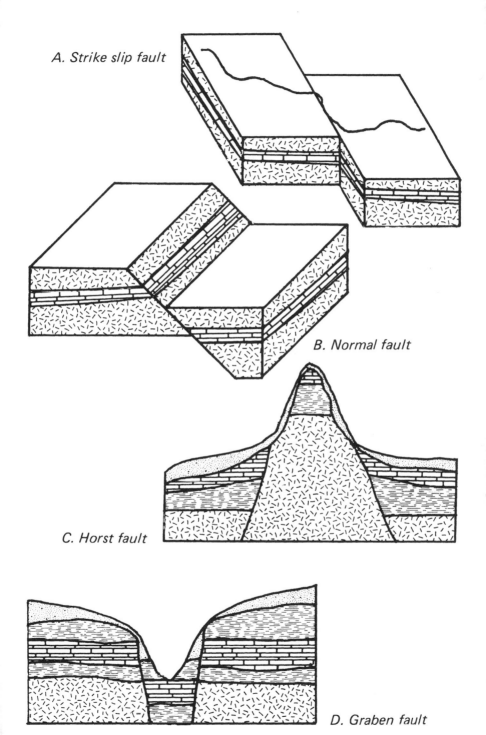

A. Strike slip fault

B. Normal fault

C. Horst fault

D. Graben fault

23

Oxidation can be caused by air but is greatly hastened by water. Minerals which contain high percentages of iron are greatly susceptible to oxidation.

Hydration is a chemical union of water with another mineral. It should not be confused with oxidation which is a different type of process. In hydration, a mineral can be altered so much that a new mineral is formed, i.e., anhydrite to gypsum.

Carbon dioxide is present in air, water and soils in great quantities. When combined with some minerals it changes their composition greatly, often creating new compounds by a process called carbonation. Also, the union of water and carbon dioxide produces carbonic acid which is even a stronger method of changing mineral compounds.

Solution is a physical weathering process which is simply the result of water dissolving the minerals and seeping downward or percolating towards the surface. This geologic phenomena is responsible for many of the rich outcrops of minerals which have been discovered. Mining geologists call it leaching and many minerally rich outcrops have been produced by some of the carrying rock being leached out while the valuable minerals remained near the surface.

The process of weathering has been going on for millions of years and continues today. Except for rapid removal of loose soils by air or water, it is seldom noticed as only minor changes in the topography are made in several normal lifetimes of man. The speed at which weathering takes place is goverened by the composition and physical conditions of the rocks being attacked, the climatic conditions and the topography of the area in which they are located.

Vulcanism, the second of the major earth forming movements is not of as great a concern to mining geology as the other forces. It is simply the name for volcanos. A volcano is an opening in the earth's surface through which magma from deep inside the earth is transported swiftly to the surface. While lava produces many gem quality minerals for the lapidarist it does not produce commercially mineable materials except for such items as pumice.

Tectonism, the third major earth forming movement is when the earth's crust is deformed by immense pressures. Usually such a movement takes place over a long time and is not perceptible to anyone but a scientist. Even when the movements are rapid, such as the sinking of Long Beach, California due to the removal of vast pools of oil beneath the city, the movement may not be detected for many years. This particular effect took place over a period of about fifty years, not even a grain of sand in the geological timer.

However, tectonic movements can occur very fast, in a matter of seconds, and when they do the result is called an earthquake.

The most important types of tectonism are epeirogeny and orogeny. Epeirogenic movements are the slow changes and take centuries and even eons to make a perciptible change in the crust of the earth.

Orogenic movements are known as mountain making movements. As a rule,

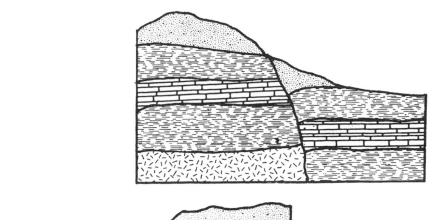

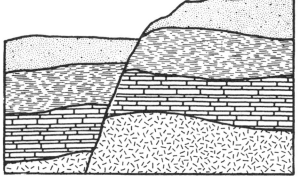

geologically speaking, they are more rapid than epeogenic changes but still their history is told in thousands of years.

The rock deformations resulting from these two movements are exceedingly important to the prospector and he will find the following defined terms in almost every scientifically written geological report of a mining district.

Warping occurs when tremendous areas of rocks are raised or lowered beneath the earth's crust. Upon cursory examination the strata would appear as a straight line parallel to the surface. Careful scientific examination would reveal that the strata dipped (was inclined) gently towards the center of the earth.

When rock strata has been subjected to extreme pressures from two or more directions it is said to *fold*. Folds can run in one direction, such as a tree bent in the wind, but a more common type is when the strata is curved up or down much as the effect of putting pressure with both hands on a flexible ruler. When pressure is applied, the ruler bends up or down.

Of this latter type of folding, an *anticline* is when the strata is folded upward with the outside curve lines pointing down. A *syncline* is created when the strata is folded downward with the curved portion pointing upward.

When diagramming folds, geologists used exactly the same terms that they use

25

in describing a mineral lode. This is called the attitude of the strata. The directional plane of the strata is the compass heading taken on a horizontal plane where the strata intersects it (not merely the surface which it outcrops as this might be inclined or curved as in the case of a mountain). The *dip* is the angle of inclination formed between the strata and the horizontal line used for the *strike.* Other descriptions of folds not often used in mining descriptions are: *monocline,* a simple fold which resembles a step and dips on only one direction; *domes,* which resemble a dome of a building with the upper part of the dome near the surface; *basins,* which resemble a basin with the lower part towards the center of the earth.

Fractures are another important part of geology to the mining fraternity. This is often how lodes are exposed, and sometimes lost in the mining process. A fracture is simply that, a well defined area where rock strata has fractured or moved.

Simple fractures are called *joints* and they normally form due to contraction due to cooling of the original rock strata. It is characterized by little or no movement of the strata in relation to the part it has fractured from.

When tremendous pressures cause considerable movement along a section of the earth's crust it is called a *fault.* Due to much publicity about earthquakes in recent years the term is much used, but seldom understood by the average person. Before describing the different types of faults it is first necessary to list the words used to describe the many parts of a fault. The rocks or strata are moved along the *fault plane* or the surfaces where the fracture takes place. The amount of movement along the fault plane is called the *displacement.* As previously mentioned describing folds, strike and dip have identical meanings when applied to faults. The surface bounding the lowest side of an inclined fault plane is known as the *footwall,* and that above is the *hanging wall.*

6

STREAM GEOLOGY

The most important geological agent (or force) known to the placer prospector is water. While most of the actions of water are pretty much self evident it is important that anyone who prospects has a complete understanding not only of the process but the words geologists use to describe various phenomena.

The whole process of how water is acquired and used is called the *hydrologic cycle* and it is a true circle. To describe it we can begin any place in the cycle. Forgetting the fact that small amounts of water are still being brought to the surface from their original pockets of formation—all the earth's water can be said to fall to the surface from clouds as rain or snow.

Once the water has reached the earth it starts on the return trip back to the clouds. By far the largest portion of the water is *evaporated* back to the atmosphere directly from the sea, rivers, lakes, moist earth or any other damp surface. The next big source of water for clouds is obtained through *transpiration*. This process has been described simply as the way plants breathe by taking water from the earth and passing it on to the air. In some areas of the earth as much as 95% of the local water is returned to the atmosphere by evaporation but the average is more in the range of 55% to 60%. The other approximately 40% is accounted for by either *runoff* or *infiltration* according to the topography, porosity and many other factors of the local soil. Runoff describes water which cannot be absorbed or evaporated immediately and begins to travel downward until it reaches the sea. Infiltration means that water which can be absorbed by any given soil until it can absorb no more.

Streams to the mining geologist include every type of moving water from narrow dry washes to muddy rivers which may be a mile across. Streams that flow only occasionally at periods when they are fed by rains are called *intermittent:* those which are fed by melting snows or have reached a depth where the water table feeds them constantly are called *permanent* streams.

One classification of streams is by their drainage patterns. *Rectangular* streams form a pattern which runs along square or almost square angles. Those that are *dendritic* in nature appear in an overview to establish a tree-like pattern and are usually found around the slopes of a mountain. *Trellis* streams are those which drain at almost right angles into the major stream.

Streams are also described according to their relationship with the topography or underlying strata. A *consequent* stream's direction flows along the original slope of the land. The direction of flow of the *subsequent* stream has been changed by fractures or a major difference in the hardness of the bedrock, usually taking the course of the softest rock. An *antecedent* stream sticks to its original course despite any radical change in the topography of the surrounding land. *Superimposed* streams have cut through softer overlying strata and now have older underlying rocks for bedrock.

The major work of streams is *erosion, transportation* and *deposition.* This work begins on an inclined surface where water falls as rain or snow, or where ice has formed. The amount of runoff for any given area is controlled by four major factors: A steep slope; the inability of the surface to absorb moisture (such as a solid rock surface; scanty or a complete lack of vegetation (as after a forest fire); and the nature of the storm (a short, very heavy rain or several days of gentle rains).

There are three major types of stream erosion that are important as geological agents: *Abrasion, solution and quarrying.*

Abrasion is the ability of a stream to wear down the sides, bedrock and boulders contained in the stream. The speed at which any given stream can abrade surrounding surfaces is governed by its *load* (the amount of material transported by the stream at any given time). Sand, pebbles and even boulders transported by the stream become cutting tools as effective over a period of time as a grinding wheel. While the stream is abrading the surfaces and carried material, boulders are also knocking together and being broken into fragments. This latter process is called *impaction*.

The second of the erosional types is called *solution*. This means that a chemical, or chemical compound has dissolved in the stream and acts chemically on rocks or minerals in the subsurface. An example would be carbonic acid, which can be created by water, air and decaying vegetation. Rocks attacked by solutions dissolved in stream water include limestone and dolomite.

The third important erosional quality of a stream is *quarrying*. This means that the force of the water literally "plucks" weakly constituted or cemented material from the banks or bedrock and begins transporting it downstream. In some cases this leaves a marked undercut in the bank or a crevice in the bedrock.

The theory of all placer geology leans heavily on the rate of stream erosion and transportation therefore it is important to understand the descriptions as they are often found in mining literature. One of the more important contributors to erosion is the volume of water that a given stream carries. Therefore larger streams, or smaller ones in flood stage, are the most active in the process of erosion, transportation and deposition.

Velocity is the speed at which the water travels and it is governed by *stream gradient* which simply means the angle at which the streams flows compared to the level surface of the earth's crust. The steeper the angle, the higher the velocity and the greater the erosional potential.

The effects of stream erosion help to govern deposition and many of the peculiarities of streams that the prospector will encounter during his explorations. Some of these are observable above the surface of the stream and others found beneath the surface.

First, to be scientifically true, all streams and valleys are formed, or try to be formed, in a "V" shape. Naturally this rarely is true in nature. The underlying bedrock virtually halts the downward erosion and the stream begins to widen, creating, at bedrock, a relatively flat bottom. The final shape of any stream bottom is determined by its discharge, velocity, nature of the load and the resistance of the bedrock. Although most of the streams prospected today appear to be relatively stable in their formation, the erosion of bedrock goes on.

Rapids and their close cousins, *waterfalls,* are two very common effects of erosion that the prospector encounters. Rapids occur when there is a very quick drop in the gradient or a rapid narrowing of the stream width. When the drop is vertical or very nearly vertical for several feet, the effect is called a waterfall. Potholes are formed when the stream water eddys or twirls in a circular motion. Its load is suspended and actually works like a grinding wheel to form a basin, or pothole, in the stream bed. When two streams flow close to one another and through soils which have differences in resistance to erosion it is not uncommon for them to meet and one stream to be diverted into the other. This phenomenon is known as *stream piracy.*

28

When streams reach maturity it merely means that the velocity is approximately balanced with the load (except at flood stage). At this period of its history it is called a *graded stream*. This type of stream includes most of the streams of modern times. However, although such a stream now cuts down only minutely, it still has the ability to erode its sides. In many cases the stream will wander from side to side in "S" patterns called *meanders*. Occasionally during flood times the water will cut across the "S" isolating a curved section into an *oxbow lake*. Finally a stream that unites and splits time after time is called a *braided* stream.

While not all of the different types of streams will concern the modern amateur prospector, it is helpful if he knows the descriptive terms for most of them. For by doing so he will be better able to understand the transportional work of streams—a phenomena equal to erosion in creating placer deposits.

Transportation is how gold is moved down a stream and the type of transportation of any stream is usually described by three types of loads.

The *dissolved load* is material carried in solution and is often described as "hard" water or "mineral" water. Material such as sediments like sand, silt and clay are called the *suspended* load. Depending upon the velocity, many streams transport larger particles like rocks and even boulders by rolling or sliding them along the stream bottom. This is called the *bed load*. A peculiar type of movement of the bed load is when a rock is moved down the stream by means of a series of jumps. This is called *saltation.*

As the speed of the stream slows down for any reason the load is slowly deposited on the bottom with the heaviest material at the bottom. This process is called *deposition* and it is the final process in the creation of a placer deposit. The major causes of the slowing of a stream are: reduced gradient; reduced volume, reduced discharge; obstacles such as trees or boulders; widening of the stream bed; rapid increase in depth or pools; overloading; freezing; and emptying into slower moving water or still water such as a lake. When deposition takes place the material is sorted as to size and weight with the biggest and heaviest being deposited first and always on the bottom.

Now that the reader has at least a speaking acquaintance with geology and the terms used to describe some of its functions he will better be able to understand the following chapters on placer deposits and lode prospecting. Only the briefest of facts were given here and for those who intend to make their prospecting more than just a weekend pastime it is strongly recommended that more advanced textbooks on geology be consulted.

7

INTRODUCTION TO PLACERS

While the scope of this book is intended to cover all types of weekend prospecting, one particular phase is by far the most popular. This is placer mining which most newcomers know only in its simplest form—gold panning. Sluicing, dredging and other more sophisticated types of placering come later, almost always as a result of gold panning.

Placer deposits are the most important type of gold recovery potentials available to the amateur today. Native gold is not only the most frequently recovered mineral, it is the easiest to find, recover and bring home. Therefore, one of the first things a person who wants to find gold should learn is the different types of placers and where they are found.

MAJOR TYPES OF PLACERS. All placers fall into two general classifications—wet or dry. Each type can be successfully worked for gold or other heavy minerals but those wet placers having sufficient water to sluice are by far the easiest to work, although not always the richest.

STREAM PLACERS. Almost all of the average weekend prospectors time will be spent in hunting for stream placers. Throughout the history of mankind streams and rivers have been the major source of placer gold.

Stream placers are aptly compared to nature's way of mining. Where they get the gold they concentrate and re-concentrate will be subjects gone into in great depth later in this book. The original sources of gold would most often be ancient stream beds through which the modern stream cut; lodes, mountain deposits, or other sources from which rain and running water will erode minute quantities, carrying the material to the stream. The gold is then transported downstream to where it is concentrated into sand bars, under rocks and other locations.

Stream placers vary widely in their distribution and types. Considerable experience is the only way to really differentiate between them but generally speaking they can be divided into about four specific types: Gulch and creek placers, and, river and gravel-plain deposits.

Gulch Placers. This type of stream placer is usually a small well defined area which may be as short as only a few hundred yards along a stream bed. It is characterized as having a steep hill immediately adjacent to the area and most of the gold was concentrated from the drainage area from these hills. The gulch placer is familiar to all as it is often the area which contains large boulders, which preclude large scale placer mining. In the early days these were often the first discovered by the pioneers and they were worked extensively since the only way to get the gold was by hand methods—usually panning but sometimes small sluicing. These placers were also the type that were most often re-worked by the Chinese miners who patiently cleaned out the last flake of gold. Today, many of them can be re-worked by modern portable dredges but there is seldom enough gold re-reconcentrated to make it a paying operation. They are good prospects for the amateur who wants to get enough gold to display and make an occasional lucky find.

Creek Placers. The creek placer begins to stretch out and represents long strips of the creek course where there are few boulders and only a few sand bars. They may extend for miles and many of them have been worked comercially. Today most of them lay idle due to the cost-return ratio of gold mining. They are excellent prospects for the amateur who has a larger dredge and lots of time. Usually such an operation consists of two or more operators and at least a six-inch dredge. While there may be rich pockets in such a placer, they cannot be counted on to sustain a long operation. The area should be extensively sampled before any long-term investment in time is made.

River Deposits. When a stream ceases to be a creek and joins other creeks to become a river the same type of extensive placer is known as a river deposit. It is characterized by fewer and fewer boulders to obstruct dredging and the gold is finer. In the higher elevations, these river deposits were often worked by diverting the river and building lengthy sluices and digging the then dry river bed down to bedrock. Today, some of the deposits are bulldozed. In the lower elevations where the river widened considerably they were worked by the mammoth dredge-line operations. While a large scale operation might fire the enthusiasm of some, it is best to remember that some of these operations worked gravel worth (by today's $500 per ounce standard) about seventy-five cents per cubic yard. Due to the fineness of the gold and its almost even distribution in a river deposit, this type of placer offers little, if any opportunity for the modern amateur prospector.

Gravel-plain Deposits. Although many writers define this type of placer differently, they are generally agreed to mean a large deposit of gravel, usually a tremendous area near the mouth of a river where it flows into plains or those where the river changes to a very low gradient in a very wide valley. They are sometimes confused with bench placers but the latter is much smaller.

Gravel-plain deposits are usually quite tremendous and easy to sample. As a result of the reliability of the tests they were worked extensively by commercial methods and some of the California deposits have produced as much as $100,000,000 worth of gold. There is little doubt that all of the gravel-plain deposits in this country have been discovered, tested, and worked if they were comercially solid. They offer virtually no opportunity for the amateur prospector.

From the full time, professional miner's standpoint most of this country's streams have been exhausted of their gold, or exhausted to the point that the cost of testing it out exceeds the selling price. But the amateur prospector has no such worries. He isn't bothered with taxes, equipment, supplies, payroll, shipping, overhead and other details that eliminate full scale mining. An amateur is only concerned with getting the gold. For that purpose stream placers are very important, since many are constantly replenishing their gold each year.

Here are a few of the more specialized placers which are quite often worked to great degrees of success by modern amateur prospectors:

POCKET PLACERS. Pocket placers are what the modern amateur prospector is usually looking for. Generally these are re-reconcentrated material found in a likely place that has been overlooked for some years. The big, extensive placers worked commercially a century ago are now mostly gone. But many areas reconcentrate gold in these locations and the first re-reconcentration takes place in the form of a pocket deposit in the most likely spot in a stream.

BENCH PLACERS are often confused with gravel-plain deposits since they are left-over concentrated material that was placered naturally during the development of a stream. While there are many types the best description is given with the meandering stream. As a stream meanders, changes course, and creates oxbow lakes, it abandons many of the earlier stream course leaving concentrated deposits some distance from the new stream course. In its later development the stream tends to straighten out and in the course of several hundred or several thousands of years there might be little evidence that the stream once meandered down the valley. The original bed of the stream in its meandering is now the soil of the floor of the valley. If the new stream has cut to any depth below these earlier bed levels and they are left high and dry and become bench placers. Another type of bench deposit is created when in its development the stream is much wider and deposits sand and gravel on either side creating a minor levee. Eventually the levee becomes high enough to stem the flood tide of the stream and as the stream matures it erodes its bed deeper. Once again the early deposited material is left high and dry and becomes a bench placer. A look at some of the illustrations accompanying this chapter will explain how bench placers are formed. Frequently in the developing stage of the stream it formed one or two altitudes of bench placers, one quite higher than the other. Early miners referred to these as "high" and "low" benches. Geologically speaking some natural consequence caused an area to allow a deposition to become permanent as the stream abandoned the altitude. At a lower level, circumstances were ideal for permanent retention of the deposit and it, too, became a bench placer.

Most of the very rich bench placers were discovered early in the history of western mining and were, due to the nearness of a stream and its constant supply of water, the original places where underground mining took place. Long flumes were built for water and the material was sluiced at first. Then as hydraulicking appeared these placers were the first to be mined in this method. Since this type of mining had all the elements for small time operations most of the bench placers were worked out in the 1860 to 1870 period. As the miners moved on to the great Tertiary deposits, the benches were abandoned as worked out.

During the depression these were the placers most worked by the individuals who roamed the hills. Once again the elements of running a small mine were present—little effort to get the rock and an ample water supply. The California

State Bureau of Mines made a survey of these miners in the 1930's and found that most were earning less than fifty cents per day.

This should not discourage the amateur prospector from looking for new bench placers as there are bound to be some that have not been discovered although they are no doubt quite small. Re-working an older, known deposit, should require a lot of thought, however, they were pretty well cleaned out during the depression and most bench placers do not replenish themselves.

FLOOD GOLD PLACERS. When gold becomes even finer than flour or other terms miners use to describe fine gold, it becomes "flood gold" to the old time placer miner. This variety of gold is so fine that it takes 200 to 1000 colors to make a milligram (there are 31,103 milligrams in a troy ounce).

Despite its fineness, flood gold is easily recovered by modern techniques. But its deposition does not really follow the expected laws of placer mining and this causes some confusion and quite often a lot of unnecessary work.

During flood times of the stream, flood gold is moved down a stream much as is silt. Due to its minuteness it can be suspended by the least action of the stream and even when the stream has returned to almost its normal state, a lot of flood gold can be suspended at or near the top of the water. Unlike the normal deposition of placer gold, flood gold can be deposited at or near the surface of sand bars and beaches. Despite its fineness, a considerable amount can be spread across the top of sand concentrated enough to be visible to the naked eye. These have been given the name "skim bar" deposits by placer miners and the name is descriptive. Skim the top of the bar and there is very little left. A more scientific name is accretion bar.

During the early days of western mining many of these deposits were discovered and worked by hand. This encouraged some to set up expensive dredge or sluicing operations which almost ended in failure, there simply wasn't any more gold further down.

One way to tell a skim deposit is that it is at the surface or buried no more than an inch or two. The other way is to remember that a skim deposit is always between normal water level and high water level. Since they occur only on bars or beach areas, there is no way to confuse them with the crevices which are often rich gold traps above the water line.

Scientifically explaining how a skim bar is formed will drive those who think they understand stream geology right up the wall—for it runs contrary to most thinking. Readers who have studied the different velocities of streams from top to bottom will be able to understand it more easily.

As a stream travels towards its base level it travels at different speeds at different places in depth and width. As it goes around a curve, tangential forces cause it to increase its speed near the outside of the bend. At the surface of the water it is slowed by air friction, at the bottom slowed even more by friction with the stream bed. While a normal observation of the surface of a stream may indicate all the water is traveling at the same speed it really isn't. These physical forces create a lot of turbulence along the inside of the curve and the water actually tends to flow backwards near the inside base line. These forces are what creates an accretion sand bar and during the process, flood gold will be deposited near the upper high portion of the bar. Since it is so light that it can easily be removed by stream action, only that which is deposited last remains.

In some areas flood gold deposits return in the same place year after year. In South America they are worked by native miners each year but no other gold has ever been found deeper in the gravel. Flood gold deposits occur many places in the United States but mostly along a 400 mile stretch of the Snake River in Wyoming.

DESERT PLACERS should not be confused with the *bajada* placers which also occur in many desert areas. The latter follows the ground rules for placer deposition but desert placers are so contradictory to established placer science that most writers give them a category all their own.

Those deposits known to the desert prospector as true desert placers sometimes break all the rules and defy scientific definition. They should neither be confused with eolian placers which are formed by the wind, or desert stream placers formed by true placer phenomena.

A true desert placer is one which was scooped out of a residual, eluvial, eolian, stream or other placer deposit and transported at a high rate of speed due to a desert downpour which creates flash floods. These gold deposits are swept up and carried down a dry wash and sometimes onto level desert land itself. As quick as the flash flood is created it dissipates itself even faster and the load carrying ability decreases so fast that its load is dumped at odd levels (geologically speaking). Heavy mineral deposits can be dropped at bedrock, at the surface, or at any place between. Here it stays until another downpour creates a new flash flood and starts the whole procedure over again.

This could explain how so many lost mine stories were created in desert areas. While it is easy to discount them as legend, the facts may be that they are true. And, that the subject of the legend was merely one of the first to learn the truth about the desert placer.

Due to the unique way in which it is deposited there is rarely anything but the small placer deposit dropped by the flash flood. It may be extremely rich, but like flood gold, any development or further exploration is usually fruitless. There just isn't anymore.

It is also not very productive to return to the same area again after cleaning out a desert placer. Flood gold will reconcentrate year after year in the same place but desert placers are lightning—they rarely strike twice in the same place.

GLACIAL PLACERS. Amateur prospectors who are not well versed in basic geology often believe that gold has been deposited in streams by glaciers. Often professional miners ascribed the source of any obscure placer to glacier action.

34

True, there are such things as glacier placers but the professional miner avoids them and so should the amateur.

As the glacier moved during ancient days, it scraped off loose debris and soil carrying it along with the travel of the ice. But since a glacier consisted of ice, and not water, it was completely inefficient in concentrating the heavier minerals. Most experts believe a glacier transporting gold or any other mineral would have tended to scatter the heavier minerals indiscriminately. Additionally, a gold bearing surface could have been scraped off and transported hundreds of miles. It has been proven that this phenomena could create a rich deposit, but it would never replenish itself and give very little evidence as to where it came from. This effect also might account for some of the rich float material in isolated areas which been found with no scientific explanations to account for it. One exception to the above description is a placer that has been deposited by the waters of a melting glacier. These would have the ability to remove and transport gold to a stream where it could be reconcentrated.

Glacier deposits which are known to have existed in this country were worked out years ago and are barren today. Since they never replenished themselves they should be greatly discounted by the serious amateur.

RESIDUAL AND ELUVIAL PLACERS are frequently contributors to the stream placer but each can be a true placer deposit in its own right entirely separate from a stream. Both types often occur in the western United States and Alaska. They are usually found in areas where water is available and are frequently worked by transporting water to a sluice box a considerable distance from a stream.

The residual placer is created by weathering and chemical action. Gold has been formed in some rock or mineral which gradually disintegrates by the persistent forces of nature. This breakdown is caused by storms, sun, winters and other types of natural occurrences, plus the chemical decay of the carrying rock.

Assuming that this process is carried out over many hundreds of years, the earth would naturally be placer mined by water and wind. Lighter material would be carried and blown away leaving heavier minerals in the remaining earth where they can be further concentrated by occasional rain storms and runoff from melting snow. Hence the name, residual placer. Preservation of this type of deposit depends upon enough water being naturally supplied to concentrate the gold of the residual placer but not enough volume or velocity to carry it to a stream or eluvial placer, therefore this type of placer is scarce in occurrence and accordingly difficult to locate without long experience, good equipment and a thorough knowledge of geology.

One of the more important things to remember about the residual placer is that unless the original ore body has completely been disintegrated by weathering, a vein will be found at the depth the chemical action has penetrated the outcrop of the deposit.

The so-called "seam placers" of the gold rush days were residual placers found in seams or cracks between rocks. To produce a considerable quantity of gold this type of placer has to be contained in such a seam or other limiting barrier and the weathering has to penetrate to a considerable depth.

An eluvial placer is generally just a continuation of a residual placer and is transitional between it and the stream placer. They are often so similar that it takes an expert to tell them apart. Weathered and disintegrated rock is washed or blown down a hillside and finally comes to rest in a depression or at the bottom of the hill. In other instances, an eluvial placer may occupy almost all of the face of a slope. Here by further action of weather and downhill slide, heavier minerals are concentrated and transported over a long period. Like the residual placer, the location is occasionally dry placered and sometimes worked by transporting water to the area. Unlike the residual, eluvial placers have no lode deposit directly beneath them, the original source being located uphill from the placer location. Eluvial placers are seldom found as the material is always transistory and heavy rains can move them downhill rapidly and in quantity. This type of deposit is similar too, and sometimes confused with, the *bajada* placer.

BAJADA PLACERS. If you have traveled in or even seen pictures of, the desert areas of the West, sometimes you have come across a mountain range with sheer cliffs, at the bottom of which the material from the side of the mountain or cliff has broken off and fallen to the base. This debris which has dropped to the ground will gradually work out from the base of the cliff and, as water and wind work on it, the material will gradually come to resemble a fan. The early Spanish pioneers called these formations, *bajadas,* and they soon discovered that if the mountain or cliff contained gold or other heavy minerals, it was naturally placered into concentrations by the weather. These deposits at the base of the cliff or mountain are named *bajada* placers. A more scientific name for this type of formation is, *talus.*

When the entire amount of placer gold produced in this country is added up, the *bajada* placer represents a very small portion of the production figures. This type of deposit, however, does create far more interest for the weekend prospector since it provides a lot of activity for desert residents who cannot travel to placer streams as often as they might want.

EOLIAN PLACERS are deposits that are formed by the wind. Wind action helps form residual, eluvial and *bajada* placers but a true eolian placer is none of these. It is most often found in desert areas where a huge mass of sand or earth carries a small amount of gold. When the strong winds occur the lighter materials are blown away leaving the heavier minerals to be concentrated. In desert areas small deposits of black sand are frequently concentrated in this manner and are often a clue to an eolian placer. Workable eolian placers, no matter how sparsely distributed the gold, are worth knowing about. They often lead to a lode somewhere nearby.

BEACH PLACERS. While there are many types of placers which have not been discussed, one last major classification should interest the weekend prospector—the beach placer. Beach placers occur all along the Pacific coast up to Alaska. The richest found were in Nome, Alaska; Oregon and northern California. A little research will prove that even the beaches of Santa Barbara produced some gold in minute quantities. Placering at the beach is done with a gold pan, a sluice or any of the co.nmon water concentration methods.

When the beach placers were first discovered it was generally assumed that the gold came from the sea. In a way of speaking this is true, but almost all of the gold originated along the coast or further back in the interior. As the beach eroded, most of the heavier material is carried out to sea where it was placered into concentrations by the action of the waves. In later years, the waves merely carried the gold-laden sand back to the beaches. While almost all weekend prospecting done on the beach today is done right on the shore line, it may be that the gold diver's next big strike will be heavy concentrations of gold in the shallow off shore zones. The coast and geodetic survey made soundings off the California and Oregon coasts a few years back and discovered black sand occurring in large concentrations. Before long, technology will probably put these deposits within the reach of the diving prospector and his portable dredge.

8

STREAM PLACER PROSPECTING

Joe Prospector is an optimist. Every time he puts his intake tube into a pile of sand he expects to take flakes and nuggets out of the sluice. Sometimes he does; sometimes he doesn't. Unfortunately, poor Joe thinks it's all a matter of luck. When the rock he placers beneath proves to be barren, he says: "Somebody beat me to it;" or, "The gold was washed out in the last flood;" or, "There isn't any gold left in this stream." Or, any other of a dozen excuses he can find to justify to himself why there was no gold where he looked.

Many times these alibis are true—but more often the reason Joe doesn't find gold in a known gold bearing stream is simply because he didn't do his homework. And when he does study the situation he finds that statements made by the writer of that book; or by his best friend who once found gold in the stream weren't true when he went to the stream. Granted, it was true when his friend or the writer told him based on their observations of the situation, but on the day Joe tried out these hints, things just didn't pan out.

The problems lie not in the intentions of the friends or writers; these are well meaning. Nor, does it necessarily lie in their facts; these are often true when they were obtained.

No one can state for sure that simply because a deposit of gold was found beneath a rock that all such rocks will have a deposit of gold beneath them in a known gold bearing stream; as anyone who has ever dredged for any length of time can tell you. They can also state with slight chance of contradiction that not all eddies, waterfalls, bedrock or any other stream abnormality will have gold—for they have all been there. But it is a mistake to believe that these failures always result from bad luck. Generally, these prospectors placer areas where they shouldn't have been in the first place. And this comes from a failure to understand some of nature's basic concepts. It can be summarized in a sentence:

To be able to determine where gold might be concentrated in a placer, one must first understand the action of the stream and the flow of water.

This is understandable because of all the sciences, the flow of liquids and gases is the least understood. Most of the empirical rules and mathematics have been developed through tests and comparisons of liquids. This is easily explained by an example. When water is contained in a closed container and at rest, there are scientific principles which govern its action and these can be predicted. Water at rest is no problem, when it starts to flow the problems begin. In closed pipes or channel-like canals, the flow can also safely be predicted with precision. However, when water flows in an open channel like a stream there are so many variables that only generalities can be stated. To compound the difficulties, a stream may have more than one scientific principle in operation within a few hundred yards of another.

I do not intend to make the reader an engineer of hydrokinetics before telling him the most likely place to look for gold. My intention is to acquaint him with a few generalities of how water does its work in average gold bearing streams so he can better find these locations. The next several paragraphs may appear formidable, but they are really quite simple. In being so simple many gold prospectors have overlooked or forgotten them. But their importance as a basis for finding any heavy mineral in a stream is utmost and the reader should study them before moving on to the more interesting parts of the actual placer locations.

The place to begin in understanding these axiomatic rules of nature is the original source of gold. This has been covered in other chapters and will only be briefly recapped here.

Before any heavy material can be transported or concentrated by a stream it must enter the stream at some place upstream from where it is concentrated and it must be of sufficient quantity to be concentrated. The place from which the gold comes is called the original source and these sources are usually veins, mineralized zones in bedrock, or permanent or preserved placers formed at an earlier geological age.

The original source can be some distance from one or both sides of the stream; or, it can be an exposed vein in the bedrock of the stream itself. As lode prospectors often discover, the original source is often a deposit that contains such a small amount of gold that it is not worth working. Sometimes the source will contain less than a milligram of gold per ton. But, over a long period of time, natural weathering agents will disintegrate the host rock and release the gold so it will be transported to the stream.

While there are many ways this transportation can take place, the most important is surface runoff—the phenomena which sets in when the soil is permeated with more water than it can support and it begins to move across the surface. The other type of runoff is called "ground runoff" and refers to the water that reaches the stream through seepage. This latter effect transports virtually no sediment to the stream although it occasionally moves some material through solution.

Gold resists transportation across the face of the earth but the resistance will be equalled eventually when the runoff reaches flood stages and begins eroding the surface of the earth. The speed at which it does this is determined by the amount of runoff and the gradient of the slopes surrounding the stream. Since the slopes of runoff are always steeper (and often very much steeper) than the gradient of the stream, gravity also plays an important part in the movement and transportation of gold. The sequence of placers in such a movement is: residual to alluvial to stream.

Once gold begins its trip of transportation on the surface of the earth it seldom stops unless obstructed. Unlike the bed load of the stream, soil near the banks of a stream is usually quite hard just below the surface and resists settling by heavier materials. The material is held near the surface and erodes easily during heavy runoff and moves along the surface. Eventually it reaches the stream at the point of entry.

The point where gold enters a stream is usually of little consequence to the prospector looking for a place to placer gold. There is seldom a quantity worth recovering. Since the entry is virtually always a result of flood conditions, and since the stream velocity is higher than that of the sheetwash or small gulley that transports the gold, the gold is very rapidly moved down the stream to where it is deposited. How far it moves downstream depends upon the velocity of the stream. This can be a very short distance, or a very long distance if the stream is moving fast. One thing is certain, as will later be explained, once the gold makes its deposition, it tends to stay where it was deposited. Very little of it ever goes much further downstream.

Four things happen after the gold gets to the stream: Transportation, deposition, retention and accumulation. Transportation refers to the travel downstream of the gold whether it is suspended in either the stream or bed load. Deposition occurs when the gold is dropped to the bed. Retention is the tendency of the gold to remain in the place where it was dropped. Once retained by the bed load the gold begins the process of accumulation and starts becoming a placer or a pocket.

When the gold enters the stream the conditions are usually flood because the contitions of entry are most often satisfied by a rainstorm. How far it is transported in this initial stage depends upon the velocity of the stream; which in turn depends upon the gradient and the volume of water produced by the storm which causes the gold to get into the stream in the first place.

The transporting power of a stream is greatly underestimated by most writers. The tables accompanying this chapter will give a fairly accurate estimate of the size of boulder that a stream can move at a given velocity. This table has been prepared from several sources and is only an educated guess based on the experiments of mining engineers. Since the velocity is only part of the effect (gravity plays a great part) there are probably no two streams which transport material at the same rate. It is enough for the prospector to study the tables and realize that the power of the stream to transport materials is considerable. The rest comes easy.

In summary, the main thing to remember about transportation is that once the gold, or heavy material, has been picked up by the stream it is transported until the stream load can no longer support it in suspension—then it is deposited. Deposition can take place near the point of entry or very far from it.

Before describing the next three natural effects of stream phenomena, I should describe the stream we will be considering to describe the scientific rules of placer formation. This hypothetical stream will be considered to be a perfect stream—like a canal—that is formed in cross section as a rectangle. No such stream exists but it is one of the few ways that the process can be described so that the average prospector can understand the procedure.

As mentioned earlier in this book, the material being transported by the stream in the water itself is called the stream load by scientists; and more informally, debris. The two terms will be used interchangeably in this text. Sand, gravel and small boulders which appear to float between the bed and surface are said to be suspended; debris which has been dissolved is in solution; and boulders which bounce along the bottom of the stream move by saltation.

When the velocity of the stream decreases so that it will no longer support its stream load, the material being transported is deposited.

Debris which makes its way to the bed from being suspended is said to be deposited by sedimentation; by precipitation from solution and when a bouncing stone stops moving, the process is called grounding.

By now you can probably visualize what is happening when a stream is at flood. Imagine looking at a cross section of the stream as though it were in a huge acquarium. The drawings on page 41 are drawn this way and will let you visualize the sands and gravels as they appear during various stages of the stream cycle.

In the first drawing the stream is at normal flood and a portion of the bed load has been eroded. The material eroded and new material is being mixed and swept downstream to deposit at the first place the velocity of the stream is reduced below the ability to suspend the material.

The next drawing shows the stream as the current velocity decreases to the point where it can no longer suspend material. All of the material has been eliminated from the drawing and only the action of the heavier materials is shown (the lighter material deposits the same way but ignore this for a moment). Those which have a higher specific gravity and those which may be of a lighter specific gravity but present a bigger cross section to the stream so more force of the velocity works on them are deposited first. This latter particle is simply a bigger pebble or boulder which may have been thin with a large cross section which lets the water pressure keep it suspended for some time.

The very first items to be deposited are boulders which were moved along the bottom by saltation. When they are grounded, they can act as a natural riffle so that gold is accumulated down stream at their base. Since saltation also takes

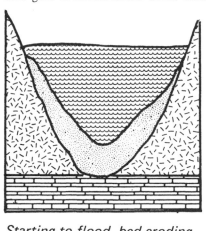

Starting to flood, bed eroding

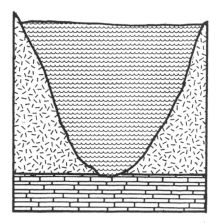

Full flood, bed eroded

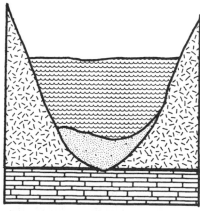

Returning to normal

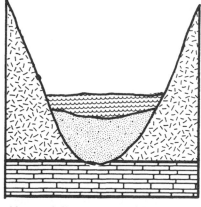

Normal flow, normal bed

41

place only when the stream erodes portions of its bed load, it is easy to see how some of the transitory pocket placers described later in this chapter are formed.

It is enough that the prospector have a firm idea of the size of material that can be moved. The bigger stones that are bounced along with the bed load grind other rocks and help to reduce them in size which also releases ore. Once gold has been released from cemented material in this manner it tends to stay where it is and further transportation is rare.

After the stream has slowed enough so that the material is being deposited at various levels along the bed load of our ideal stream, it is then retained. This merely means that the gold is covered with sand and gravel which serves as an effective cover protecting it from further transportation by the stream's water.

When this natural process of retaining gold is understood, it is at once evident why gold tends to stay where it was originally deposited. There are two major reasons why this is true.

First, the protective covering of sand and gravel must be eroded by the stream before the gold can be transported. This means that the stream must reach the same flood stage it had when the gold was originally deposited. Since most streams reach flood stage at regular intervals during an annual cycle, it is unlikely that gold deposited will be eroded before a year passes.

If gold stayed at the depth it was deposited, much of it would be picked up and transported each year as the stream reached its level of deposition during flood stage. However, there is one final process which takes place immediately after the gold has been deposited which almost insures that it is highly unlikely that any gold deposited during flood stage will ever be eroded again unless a flood of greater magnitude occurs.

This process is actually settling although it is often referred to as accumulation—it is actually a process of both.

Some material contained in the bed load of a stream generally resembles that of the bank to a certain degree. There are some important differences. First, it has been ground and reground until its consistency is far finer than that of the shore. Second, it is always penetrated by water and has a much higher viscosity (simply stated viscosity is a measure of the slipperiness of anything). This moisture content lubricates the gravel and allows heavier material to pass to the bottom.

Besides its high specific gravity, and the ideal conditions for settling contained in its environment, there is one more factor that hastens the settling of gold.

This is the normal action of the bed load; that accumulation of debris that is contained in the space between bedrock and the bottom of the stream. The gold that was deposited by the stream load during the decreasing velocity of the post-flood stage is now distributed more or less regularly at different depths.

The bed load is also being transported down stream. In other words, that portion of the bed between bedrock and the bottom of the stream is also being moved downstream. This could be termed a sand flow, only in slow motion—very slow motion. This movement is due partly to gravity and partly due to the force and weight of the water in the stream. This movement includes not only the sand and gravel, but the larger boulders that are not firmly attached to bedrock. As the bed moves down the stream many actions are taking place.

The movement down towards sea level presses the grains of gravel and sand against each other allowing releases of pressure which help to speed the process of the gold toward bedrock. The heavier minerals also tend to migrate downstream and their trip to bedrock takes a curved path.

Eventually, after a long period of time, the gold reaches bedrock where it can sink no further and begins moving downstream parallel to the bedrock. If, in these travels the gold crosses a crevice, it will generally gravitate to the bottom of the crevice where only the most violent of flood stages can remove it for further transportation.

Another peculiar effect of settling is the accumulation of paystreaks. As it makes its downward trip, the speed of the settling is slower and slower as sands become more compressed and the lubrication of the water less. This means that as some pieces begin slowing down, other flakes of gold catch up to them and the two join to settle together. If the trip is long, and the amount of flakes settling many, a large number of gold flakes can combine to settle together. This is called accumulation (although the term also means accumulating in a pocket placer as will be explained later) and these accumulations were the pay streaks the operators of the old bucket dredges used to recover with regularity when these machines were running. There are many ways these paystreaks can be accumulated in placers. They can be a small, concentrated area; they can gather over a large area; or they can accumulate over an entire layer of the placer.

With this explanation of how gold enters a stream; is transported, deposited, retained and accumulated; the modern day prospector has a big edge on his predecessors. Today, we have the observations of hundreds of years of mining and scientific investigation to draw on and by using what others have learned we can more accurately judge the stream. The old timer called it "reading the stream" and this is a good description. Today, you are not going to have his years of experience, but by remembering a few rules you'll have enough information to get started.

If you think finding gold is going to be easy, forget it. Most of the information that has been given amateur prospectors is misleading. It has been based on personal experience but unfortunately some of the writers and advisers jumped to the conclusion that just because they found gold in a certain place, all places like this would turn up gold—it just ain't so.

The stream described in the first part of this chapter was an ideal stream. One that we chose because it could be sliced in the middle and observed to show exactly what was happening during the different processes of gold reaching bedrock. Few streams have such a section in very great length. If they did the gold would be so evenly distributed that it would take centuries for enough of it to be deposited so that it could be dredged out. Also, during this time the gold might very well be all washed out by a major flood leaving the area to start accumulating gold again.

Fortunately for the modern prospector, streams meander, bend, twist, narrow, drop over cliffs, and do a thousand other unique things that separates each stream from the other. When this happens the stream creates a natural riffle or gold trap and this is what the modern prospector wants whether he is dredging, sluicing, or panning. These are the pocket placers and the supply of them will probably never be exhausted. Many are replenishing themselves and others are being created as a rock moves to a different place in the stream or as the stream erodes a new curve. The process of replenishing may take one year, five years, ten years or twenty-five years in the process. While I said there are probably plenty of them left, you'll have to find one that someone has overlooked for one to twenty-five years or more. But when you do, the wait will have been worth it for nothing can match the thrill of finding your own gold.

There is no way a person can be taught how to read a stream in a book—but the following information will make it possible for you to teach yourself.

First look at your prospective stream and try to imagine it as it was at flood stage. Then use the information in the first part of this chapter and try to imagine what happened when the stream began to reduce in velocity. Lastly look at your stream as it is during your visit and try to put the three stages (flood, receding and normal) together in your mind. The rest will come easy and the chances are good you'll come up with several good prospects for placering.

When you first read your stream look for two things—where the water slows and where the water is in turbulence.

At places where the water is turbulent it breaks up the gravel and erodes the side or bottom of the stream. This separates gold from its carrying matrix and allows it to settle. It may be that the turbulence is strong enough only to carry away lighter material or it may be that it is stirring up everything and transporting it downstream. But below the turbulent area there will be slower water and this is a good area for deposition.

Wherever water slows is a good place to investigate for gold accumulations. There are several generalities of where water slows and these should always be looked into. Water slows when it enters a wider portion of the stream, it also slows on the sides of a bend and immediately after leaving a waterfall or rapids. Some of these sections of the stream are very obvious and easy to spot. Some places where water slows are very subtle and need closer examination to determine if they actually deposit gold. Some, like bends, waterfalls and rapids are so important they will be discussed in the pocket placer section which follows.

Before going too far it might be a good idea to go a little deeper into what a pocket placer is. The hundreds of years available for the great placers to form are gone forever. True, there are localities found every year where several ounces can

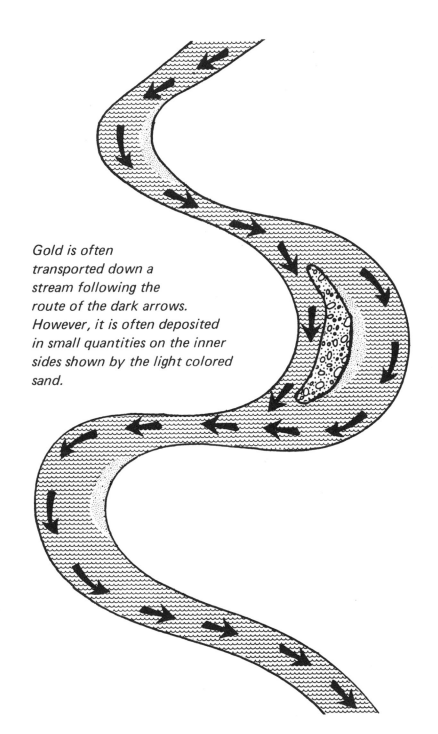

Gold is often
transported down a
stream following the
route of the dark arrows.
However, it is often deposited
in small quantities on the inner
sides shown by the light colored
sand.

be recovered in a period of a few days but the days of the bonanza placer are probably gone forever—at least in the known gold producing areas of the United States. Today's amateur prospector will work for days just to recover an ounce or two. Nearly all of the gold being recovered today by a weekender is found in a pocket placer.

A pocket placer is, by definition, a small deposit usually limited in quantity to an ounce or less. It is formed by a natural riffle in the stream proper or along the banks.

Pocket placers are fairly recent in age and most were formed within the period of a few years. There are many exceptions to this, and pocket placers retained in bedrock crevices may be more than a century old. Due to its comparative newness, a pocket placer is more susceptible to the erosion of the stream at flood time than most older placers. During this period about four things can happen. If the flood does not erode down to the depth of the placer it is retained and nothing happens. If the flood does reach the depth where the placer is located it can be washed away completely to either reform as a new pocket placer some place else or to be dissipated completely so that discovery or recovery is improbable. This usually happens when the natural riffle is destroyed or displaced during the storm. This often occurs when a big rock moves several feet by saltation.

The natural riffle can be of unusual construction so that it permits the agitation—but not the transportation of the gold it traps. If new gold has entered up stream it will probably be trapped to increase the quantity of the gold in the pocket placer. This often happens in a pot hole, eddy or waterfall.

Lastly, the entire concentration of gold may be washed out completely only to be replaced by a new quantity of gold from up stream deposits or new gold entering the stream. Float gold is an example of this type of effect and it is often true of individual sand bars.

Pocket placers can be divided into three general types: Permanent, transitory and temporary. Each type and the causes that create that particular pocket will be described in length but the types themselves should be further explained first.

Permanent pocket placers are those found at bedrock. These are generally the oldest of modern day placers and they are seldom younger than twenty-five years and often a century or more older. They are the most easily preserved, for the stream seldom reaches bedrock during flood stage. They often prove to be the richest, returning more gold than any other type. The biggest drawback to this type of placer is the quantity of barren sand that must be removed to find the gold. Bedrock is often sixty or more feet deep and is seldom nearer than twenty-five feet. Amateurs seldom attempt a large bedrock placer unless thay have three or more in the party and a dredge no smaller than six inches.

Transitory pocket placers are those found between bedrock and the surface of the bed load. They are the type most often found with the modern mining equipment available to the amateur prospector today. They are usually from six to ten feet deep and have been formed within the last few years. They are seldom more than ten years old but this varies according to the immediate past weather pattern. They are almost always concentrated by a natural riffle like a rock, car body, etc. They are frequently found in areas where the bedrock is close to the surface and pot holes or other depressions have been created. In many cases, the transitory pocket placer is still settling but the accumulation of the pay streak has been created by a natural riffle which brings the gold together so that it can settle as a unit.

Temporary pocket placers are the accumulations found above the current water line but below the high water mark. They are very recent in age most having been formed during the last flood stage of the stream. They are usually found in a natural riffle, depression or crevice and except for special cases like nuggets wedged tightly between rocks, they seldom remain long. Some temporary placers are so fragile they are washed back to the stream by the next rain. They are easy to find by exploring banks and this makes them hard to find if you are looking for them. Most are discovered by blind luck and by persons who do not placer but are merely "out for the day." They cannot be relied on to be anywhere as most are accidents of the stream's transportation so few are formed.

Those just getting started in prospecting are always looking for specific places to find gold. This not a bad quest but to tell the truth there are so many specific places that it would take a book to list them all and the day it was published there would be more to add to the list. Too many people rely on such lists without ever realizing they are really lists of specific places where gold was once found. The only true list can be a short one describing those places where gold is most likely to accumulate applied to specific abnormalities in a stream course. When you start applying the list you will find that these suggestions will apply to many situations and once in the field you will have to rely on your own judgment to find the best prospect to try out.

In describing them I will stick to the three basic types of pocket placers and describe several different places where gold might be expected to be found. You will notice that some of the classifications are listed in more than one general classification. This is to remind the reader that if gold is found under a waterfall, for instance, it could occur either at bedrock or between bedrock and the surface.

PLACERS FOUND AT BEDROCK (Almost always permanent)

Crevices. Of all the types of gold deposits that might be profitable, bedrock crevices have to be near the best. A crevice for this purpose can be defined as a split, tear or crack in the bedrock. It can be only a few inches deep and several feet long. Crevices cannot be predicted. Some run parallel to the stream current, some at right angles and there are variations of angle between these descriptions. Regardless of the way they run to the current, all are possibly the most efficient of the stream's natural riffles. Crevices are formed in many ways but two of the most prevalent are the contact point between where two geological formations meet and the splitting of the bedrock as it is eroded. There is one sure way to locate a bedrock crevice from above water level. This is to watch the sides of the stream to see if a split in the rocks is continued into the stream. These can be cleaned out from the side of the stream down to bedrock and often followed for some distance without having to remove a lot of extra gravel.

The more productive crevices are often hidden well beneath a thick bedload and never reached by the average dredge operator unless he is removing the bed load because investigation shows it has enough gold to warrant dredging. Often when a dredger is working a large area of a stream to bedrock he does just that and fails to make a close investigation of the bedrock itself when he reaches it.

When going to bedrock it is always a good idea to spend some extra time with the intake nozzle a few inches away from the bedrock, even though it might appear that there is really no crevice there. Bedrock often has minute depressions and minor crevices which appear to be only a scratch in the surface. These can easily be cleaned out by vacuuming the bottom. It is well to remember that crevice an eighth-of-an-inch wide and a half-inch deep and a foot long might have more gold in it than several yards of bed gravel.

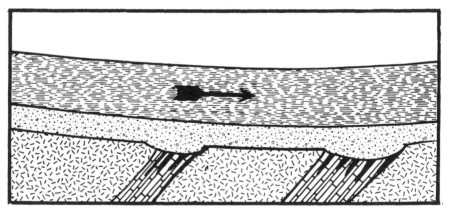

Crevices in a stream bed are one of the best bets for Placer gold. They are rarely disturbed by the strongest floods.

Surface to Bedrock Boulders. One of the most overlooked possibilities of the bedrock pocket placer is the boulder that is so huge in size that it protrudes from the bank or center of a stream and extends down to bedrock. The fellow who will always stop to clean out a boulder which does not extend to bedrock will often pass these up.

While no general rules can be laid down as to which might be profitable, each should be analyzed as to its shape and the way water might be passing it at flood time. The general principles are the same, analyze the huge boulder as though it were smaller and try to determine where gold might be deposited. This is rather

48

difficult since many of these giant boulders extend down some depth and considerable overburden must be removed to get to paydirt. Also many of them are located in the middle of a stream and work is difficult, if not impossible most of the time. Either is a good prospect, however, and there is one thing a person who has access to a stream on more or less regular basis might do. This type of bedrock placer is seldom worked for the reasons just mentioned. However, when the stream is at its lowest point in velocity, much more of the rock is often exposed and working conditions become much easier. Exploration of the huge boulder might wait until this point in the season and then be worked without having to fight strong currents or remove a large amount of unproductive gravel.

Faults. Although often hidden, faults are usually exposed on both sides of the stream and the underwater characteristics can often be predicted with reasonable accuracy. The type of fault the prospector is looking for here is not of mountain building stature but usually one of those that geologists find too insignificant to map. Underwater erosion of a fault is usually far less than that on the nearby exposed shoreline and although exaggerated in the illustration with this chapter a good idea can be formed as to how they act as a natural riffle. The gold trapping ability of a fault is usually as good if the lower portion is up stream or down stream so no matter how it looks on the surface examination, a fault is well worth investigating.

Dikes or outcrops. Dikes or outcrops in a stream seldom resemble the nice sharp outlines that mining illustrations give them. They have been eroded by stream gravels for so many years they usually look like a huge boulder resting on bedrock. Like the boulder they should be examined as a riffle to determine whether the most likely point of deposition is up or down stream.

Potholes. Potholes are formed either at a spot where the bedrock is softer, or

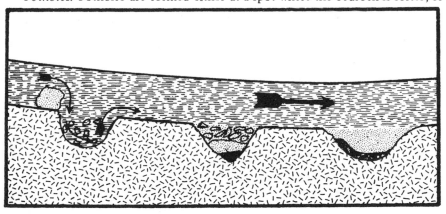

Potholes are many and varied. First pothole on right receives considerable turblence allowing gold to be pulverized and washed out. Other two are well cemented.

49

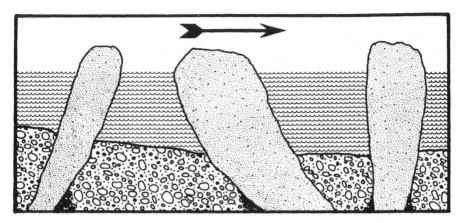

Boulders extending some distance into the stream bed are always worth investigating. Gold can accumulate downstream, upstream or on both sides, so take nothing for granted.

where it receives more abrasion and erosion. Contrary to popular opinion, they can be formed where the water does not eddy. Sometimes a small portion of the bedrock is so soft compared to the surrounding material that it is simply scooped out. A good example is where there was once a mineral deposit that was softer than the carrying rock. Other times a huge rock can tumble, ripping out a portion of the bedrock.

Also, potholes can be very small, or very large, sometimes even twenty-five feet in diameter in large rivers. And, eddies do not always cause potholes; eddies are sometimes caused by potholes.

Like all geological structures, potholes have various stages of development and react differently at each stage of the river's action. When they are relatively new and small, material tends to slide over them abrading the hole larger and depositing little or no heavier material. As they become bigger, heavier material is trapped within them and retained by sands which pack the pothole full and little more abrasion occurs to increase the size of the hole. It is at this stage of development that the pothole becomes interesting to the gold dredger and a small one can contain a good deal of gold.

Unfortunately potholes are usually buried beneath the bed load with little surface indication that they exist. They are most often found by the individual who is dredging to bedrock and some depressions are so slight that the dredger may clean them out without knowing that the pothole was responsible for most of the gold he recovered that day.

Where the water slows. All types of pocket placers are formed where the water slows. In searching for bedrock placers it is well to go back to the beginning of this chapter and remember the several speeds of stream water. A bedrock placer often cannot be found by the way the water slows at normal level. Try imagining the stream at any point when it was in flood stage by trying to determine how high the water was then. This can often be done by observing the water line in any particular area. Then compare this observation with the current level and you will get a good idea where the gold was deposited. Be sure to remember that we are now talking about bedrock placers and not only deposition but retention and accumulation are the most important factors. Here are a few of the more important places to look.

Curves or bends. Despite what you may have read many places, the inside curve of a bend is not always the place where the water slows. This is so crucial to not wasting time in prospecting that a series of illustrations accompany this chapter to show how a curve is usually scooped out on one side where the water flows deep and on the other where it flows shallow. Too many writers have given the impression that all you have to do is locate a stream and follow the inside bends and that gold flakes and nuggets will follow a well defined path down the stream. Look at the illustrations before you form any hard opinions that you will never break.

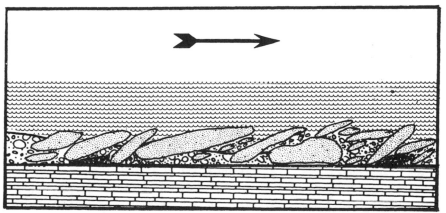

Rocks "shingle" on a stream bed near bedrock. An accumulation can be a good gold trap and many have given up a considerable amount of placer gold.

Rapids. A rapids in a stream is generally a place that was once a waterfall. This is the final stage of the water fall and many of the rules of both apply to each. The area immediately down stream from a rapids is almost always slower water but it cannot be guaranteed to have a gold deposit. Many rapids in a stream run in sequence and are characterized only by white water in the points where rocks break the surface. The velocity over such a series is pretty much the same and any gold will tend to be washed out instead of deposited. An ideal rapids which could be expected to have a gold deposit below it is one where the width of the stream immediately widens or deepens causing the velocity to slow considerably. Remember that we are talking here about bedrock deposits so the

51

analysis must be made as though the stream was in flood. At this stage of velocity, the chances are good that the water covered the rocks and was at a much higher velocity. The chances are good that gold was simply transported past the rapids area. There is a possibility that the rocks acted as a natural riffle and accumulated some gold, but what happens to it later is covered in the next section.

Waterfalls. There are waterfalls and there are waterfalls. We will not be concerned here with the very small ones. The term small also includes the velocity, so there are those in this paragraph and those in the next section. Here we are concerned with waterfalls of at least more than moderate height and velocity. Generally speaking they would be of stream width, almost vertical descent (the water must drop, not rush down a chute) and year round operation. To create a bedrock placer the waterfall must be in operation even during flood time.

Waterfalls like this are ideal for transportation and almost all gold washed over them is transported with no deposition taking place near the bottom. At the bottom there is a churning effect and most gold which makes its way here doesn't stay around long enough to be even ground up. It is swept up into the current of the stream.

However, a major waterfall almost always empties in an area that is wider than the average width of the stream just above the waterfalls and for some distance below it. This is often a straight stretch of stream which acts almost like the ideal stream used to describe the four processes that occur after gold enters a stream. Gold deposited in this area will almost classically follow those rules and what is created here is often a major placer with a tremendous amount of gravel that is fairly equally deposited with gold over a large area near bedrock. Such an area is worth sampling and often worth dredging out.

Sand bars. Sand bars are the most romantic part of gold prospecting and that they are well worth working is easily proved by some of the most famous place names of the West. They can be formed in many ways but the most common is by the same methods that gold is deposited. A sand bar always means that heavier materials are accumulating in a specified area; if the less dense material has been deposited so will gold if there is any to be had. Unfortunately, sand bars are an obvious feature of a stream and due to their popularization in fact and fiction, usually the first place a newcomer dredges. Due to this most of the sand bars in gold country are worked every year and barren due not to the fact they do not accumulate gold, but that they are cleaned out too often to let an accumulation worth dredging accumulate. For those who want to try their luck, several illustrations show how they are formed and the best spots to prospect.

Where streams widen. Many writers have always stated as an axiom that any place where a stream widens is an ideal place to dredge for gold. But what they failed to state is that this is almost always a bedrock type of operation and to be a bedrock placer of note the area must include not hundreds of yards, but more often thousands of square yards worth working. Add to this the fact that once

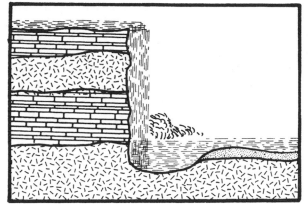

Waterfalls with considerable turbulence usually pulverize and wash out gold.

Waterfalls which have stream bed sand beneath them almost always concentrate gold.

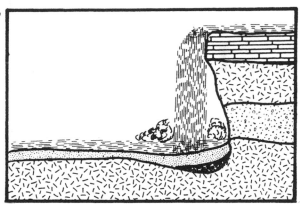

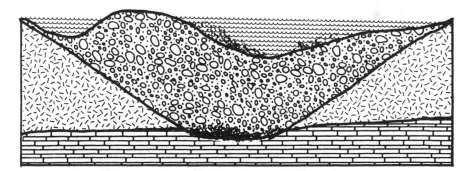

Sand bars often have a recent accumulation along the sides, but to get any quantity of gold, the best solution is to go to bedrock.

the stream widens into such an area, the gold is so widely distributed that only a major operation with something like a bucket dredge can be successful. The amount of gravel to be removed is staggering. For the amateur prospector a bedrock deposit where the stream widens is almost out of the question. Yet, in the smaller confines of the mountain stream it is always wise to try to determine where a stream widens and slows at flood time for this leads to creation of pocket placers—especially the type found above bedrock.

Down stream from a tributary stream. A tributary stream usually has a slightly steeper gradient than the stream into which it empties. This means that it has a higher velocity and can transport more gold due to its speed and the effects of more gravity. As a result, more gold entering the tributary is transported to the main stream than is transported by the main stream. This means that very soon after the gold enters the main stream it is going to be deposited and usually retained near the mouth of the tributary. This has given rise to the old prospectors "right hand rule." Simply stated it says that if a gold bearing stream empties into another, gold will be found near the right bank down the main stream.
stream.

What the old timer said was true, but he forgot to say just how far down stream the gold would be found. This depends upon the velocity of the tributary compared to the velocity of the main stream. Only experience with any two such streams will say how far with any accuracy, but it can be relied on that there will be gold downstream at some distance. Both bedrock and near bedrock pocket placers can be found this way and it is a good place to prospect.

Eddies. Eddies are one of the most misunderstood phenomena of stream peculiariteis. To get an idea of how they concentrate gold put a spoon in a glass half full of water. Now stir the water and remove the spoon while the water is still swirling. Notice that the water has risen on the outside of the glass while a distinct depression (or vortex) is in the center. This is how water in an eddy works.

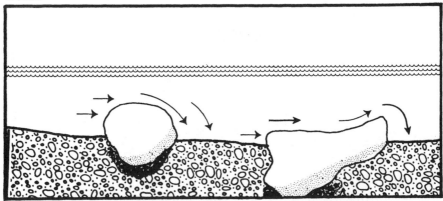

Deposits beneath boulders not extending to bedrock are usually recent and generally consist mainly of black sand.

As the water is pushed out and up, so is the suspended material. Naturally some of the suspended material is heavier and tends to stay nearer the center of the eddy.

When the stream is in flood the eddy speeds up and often stops its surface motion as the water level rises and covers the reason for the eddy in the first place. Although it is no longer visible from the surface the eddy may still swirl below water creating its whirlpool effects.

During flood times the eddy is fed suspended material by the velocity of the stream. This material is spun around with the lighter material being cast to the outside where it is captured by the stream and carried off. The heavier material tends to stay near the vortex of the eddy. As the stream abates from flood stage the eddy begins slowing down and the heavier material is deposited where it is retained and accumulated. Take a look at the drawing illustrating what happens in a riffle (in a later chapter) and you'll get a good idea of how an eddy works when it is vertical, not horizontal.

In the stream there are two types of eddies—pressure and suction. The pressure eddy is formed when water of the stream pushes against a natural or artifical obstruction. The suction eddy forms when the stream pressure passes such an obstacle faster than the water can fill up the space. Either type of eddy can be a bedrock placer and those which can be well defined are often productive when worked to bedrock.

PLACERS FOUND BETWEEN BEDROCK AND BEDLOAD SURFACE

This type of placer is usually found at a depth of from three to ten feet, and it is transitory in nature. It has been accumulated recently and if not recovered or washed out will eventually work its way to bedrock.

Logs. Logs are one of the least likely places to find a transitory placer. However, I am bringing them to the mind of the reader since logs, especially in the West, tend to range on the big side. What we normally consider a log, is generally about six inches in diameter, but many western streams capture logs from timbering operations and these can get to be three feet in diameter with lengths of twenty to forty feet. Once such a log is firmly implanted in the bed load it tends to remain and becomes water logged. Often they will split creating a crevice and these can trap gold, especially under the bed load surface level. While not the most exciting prospect found in the stream there are logs which will stand investigation.

Rocks and boulders. Rocks and boulders are one of the most lucrative of all pocket placers. To be worth investigating, either by panning or dredging, it should be of a size which is big enough to be more or less permanently placed in the bed load, but not big enough to settle to the bottom. All rocks are temporary obstructions to the stream but many of this size will remain in place for hundreds of years until a flood of unusual proportions erodes the bed load beneath them and transports the boulder by saltation to a new, lower location.

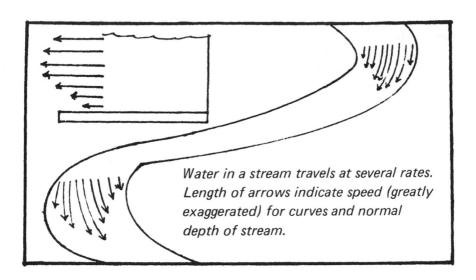

Water in a stream travels at several rates. Length of arrows indicate speed (greatly exaggerated) for curves and normal depth of stream.

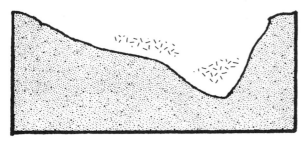

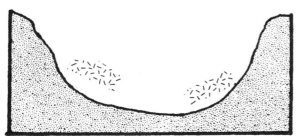

Depending upon the countour at the bottom of a stream, the greatest area of turbulence would be found at the areas indicated by the shading.

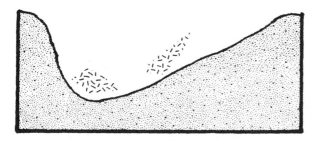

There is no general rule as to the size of a boulder which should bear investigation but it should protrude above the normal level of the stream and extend down for no less than six to eight feet. There is also no general rule that states definitely where gold will be trapped, but generally it is on the down stream side. Many amateurs have their own secret rock which they clean out from time to time.

There is a danger in digging out a boulder in a stream. Remember that there is always a pressure from the current and if you remove enough sand and gravel down stream there is always the possibility that the boulder may roll over on the operator due to an unusual shape. A good rule here is to never put yourself into a situation where you could not make a safe and hasty retreat if the rock moves. And, if the rock makes even the slightest movement, get out of the way quick. If this should happen it is a good idea to put some additional pressure on the other side by pushing to see if the rock is really loose. Never take chances with an odd-shaped, loose boulder. A stone 3x3x6-feet in a retangular shape could weigh up to 10,000 pounds and there is no way you can lift it off of you if it tumbles.

Despite the inherent danger, most rocks can be judged safe or unsafe from the surface and they are a favorite with panners and dredgers. The secret is getting deep enough to find the paystreak, and get this gravel to your pan or dredge. This is a little more difficult in panning as the gravel tends to separate quickly and mix with the water. It is a good idea to gather pay streak gravel in the pan at the lowest level and rotate the pan a few times under water before bringing it to the surface. Bringing the gravel up with a shovel always results in considerable loss and probably much gold.

Panners seldom work at depths more than four feet but dredgers with a snorkle can easily handle six feet. With the dredge it is a good idea to sweep an area wider than is necessary to make sure none of the gold value is getting away.

Rapids. Potholes and other depressions are often formed immediately below a rapids and quite often thay have enough depth to temporarily retain a small pocket placer. It works like this: As the stream decreases in velocity, the rocks of the rapids have gold deposited in them which they cannot retain. This gold is not settled but tends to remain in a state of agitation at or very near the surface. Minor changes in velocity can transport the gold down the rapids where it is deposited near the mouth of the rapids. Here the gold is retained by a thin layer of gravel and begins to settle. Settling and accumulation seldom goes very much further than the next flood stage when the gold is washed out to be deposited further down stream. A pocket placer of this nature can usually be classed as an annual accumulation to disappear each year and reform shortly thereafter. If you find one of these it is well to determine when the maximum accumulation is formed and clean it out once a year. There is little chance that it will get much bigger in future floods.

Waterfalls. Waterfalls differ from rapids in that the water has an almost vertical descent—those considered for transitory placers differ from bedrock placers in that they are not as big. From the placer prospectors standpoint there are two types of waterfalls to be considered. One is the type where the rock at the base is as hard as the surrounding material and very little abrasion takes place so that all gravels deposited here are soon washed away. This type of waterfall is of little consequence and heavier material contained in the gravel must be searched for downstream at a more opportune location.

The other type of small waterfall is one where the rock at the base is of a softer material than that surrounding it. In this case the waterfall will erode a depression which will retain sand. After the base has matured it will retain as much as it allows to flow out and since most of the material is being agitated in turbulence, heavier materials tend to settle and stay longer than the lighter material. After a period, such a waterfall will have a deposit of heavier materials. Many very rich placer pockets have been recovered in such waterfalls, but they are few and far between. Those who know where they are return year after year—and almost always with good results. Ilustrations on page 53 will give a good idea of the two types.

Car bodies. Car bodies are mentioned only as a description of all the man made items which often find their way into stream beds. Any exposed metal item is rarely a point to start looking for accumulated gold. However, iron and iron products settle and accumulate much the way that gold does and often in the same general area. If you start dredging up nuts and bolts from the ten foot level—or if you uncover an old car body down deep, it is a good idea to clean out a large area immediately surrounding the area where these items were recovered.

Tree roots and vegetation of any kind. One of the unusual spots along the side of a stream and below the high water mark where gold is often found is alongside the roots of trees and grasses growing there. Unusual because this spot defies all the rules of deposition. Roots have tiny hairs growing from them and these trap small particles of gold and hold them. It is virtually impossible for them to escape or even settle. Situations like this are best ascertained by visual inspection or panning small quantities. The usual procedure is to pull up a few clumps of grass or to dig out enough soil with the hands to fill a pan. In some cases the area around a tree can be dredged. There are some hazards, however. Land owners, land management officials and others take a dim view of removing the soil around a tree and besides it might fall on you. Use caution and good sense if you find a deposit like this.

Shingling rocks. Rocks at the bottom of a stream have a tendency to shingle, or mesh one upon the other. The situation is shown in illustration 0-0. This has always been true and cases of shingling in the drift mines of the Tertiary rivers of California were used by the miners to determine which way the stream flowed in ancient times.

Shingling is much like the rock riffles which were used in the hey day of California placer mining and are excellent gold traps. Shingling can occur near the bed load surface but is often found well settled into the bed. Wherever a stretch of shingled rocks is found, it is a good idea to investigate them by further placering.

Rocks in well defined groups. Many places in streams of all sizes, groups of rocks are found in close proximity to each other. This should be viewed as a natural riffle and bears more than a casual inspection. The sand around the rocks can be dug out by hand and panned but if there are many rocks and they are quite close together the problem is that the holes fill up too quickly for a panning operation. Sometimes a diversion dam can be built but more often a smaller dredge works wonders.

If the rocks are close enough together the possiblity that the most gold is deposited near the first few rocks on the up stream side is just as strong as finding the most gold accumulated in the first few riffles of a sluice. For the panner who has limited time and resources this is the best way to begin. A dredger should begin at the end of the group downstream since he has to deposit his debris behind him. Also there is the possibility of loosening the rocks and they will move down stream making the operation difficult if not impossible.

POCKET PLACERS ABOVE THE NORMAL SURFACE OF THE STREAM

These are the temporary placers that seldom last from year to year. The location of this type of pocket is limited only by the imagination of the person looking for them. Many years ago there were many individuals who made a meager living by "crevicing" as it was known then. This meant they had special tools to extract small amounts from crevices along stream beds and spent their summers doing just that. They rarely panned the stream except when the crevices played out. Mostly these were old timers who had been in the areas they creviced for many years and their intimate knowledge of the streams made them successful to a degree. As the old timers died out, or perhaps it was due to the development of the modern surface dredge, this activity died out. Today the search for placers below high water lines and the normal creek level are only casual occurrences. Sometimes they are done with a metal detector but more often it is just a visual inspection of a likely looking spot. Only three common sites will be mentioned here, but there are dozens of others.

Crevices. Splits occur along the side of the stream with just the same regularity and frequency that they do beneath the bed load. When the stream is in flood, internal pressures or accidents of suspension will push flakes of gold to the outer limits of the stream bed. Often the stream reaches a high point and recedes so quickly that this gold is dumped and often it is washed down into a crevice. Normally speaking, a crevice of this nature contains little gold, but occasionally one will have enough to excite some attention.

Where two rocks meet. Rocks alongside a creek are often piled (sometimes this is the result of an earlier mining operation) and there are sometimes deep spaces between them which allow sand to accumulate but not escape. Unfortunately nine times out of ten it is just that—sand, but the tenth time often turns up a little gold. Usually these are quite deep and many ingenious tools have been developed to spoon out the sand so it can be panned.

Underneath rocks, natural items, logs, etc. The instances you will hear of something being found underneath rocks or other items that have been turned over in a stream are usually not true or exaggerated out of proportion. If one deep crevice above water has a small portion of gold, the percentage of finds for rocks and other items must be at least one in a thousand.

In summing up the potential of stream pocket placers, only a few of the many hundreds of variations have been mentioned. Rather than go in the field with merely the thought of looking for a rock; or looking for a waterfall; the potential prospector is well advised to digest carefully the principles laid out in the first portion of this chapter. When you understand the way pocket placers are formed, then you will be able to judge the specific loactions far better than this book—or any other book for that matter—will ever be able to tell you.

9

AN INTRODUCTION TO DREDGES

There are several types of modern portable dredges. Each has a use and in certain situations one will work better than another. Basically, they all work on the same principle with only a few variations of design which does not alter the overall concept.

Dredges are first classified by where the sluice is located: either above the surface of the stream or under the surface. Next, they are classified as to where the ejector is located either above, near or below the surface and finally, by whether the ejector is fixed or movable.

When the sluice is located above the surface the dredge is called a surface

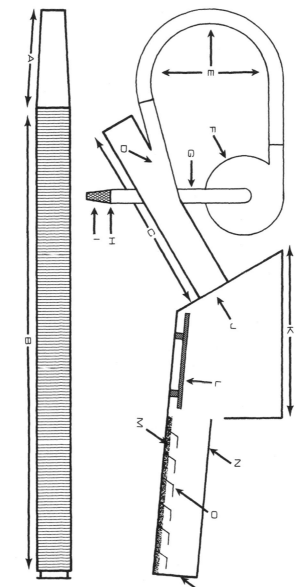

A. Suction nozzle
B. Suction hose
C. Eductor (Ejector)
D. High pressure orifice

E. High pressure hose
F. Centrifugal pump
G. Suction hose
H. Foot valve

I. Strainer
J. Ejector outlet orifice
K. Pressure Box
L. Classifier (Grizzly)

M. Mat or rug
N. Sluice
O. Riffle
P. Sluice

61

dredge; when the sluice is located between the surface and bottom, it is called an underwater or submersible dredge. Illustrations on page 63 show some of the many configurations now being made by most manufacturers.

Before describing and further classifying different types of dredges let's go into the background of the dredge and examine the principles on which it works.

First of all, modern dredges do not operate on the "Venturi" principle, which is a physical theorem given by Bernoulli who first proposed the fact that the pressure of a fluid decreased when the velocity increased, which is a restatement of Newton's principle of conservation. This is how the flow of fluids is gauged.

We will discuss all that later. First let's look at the granddaddy of the dredge—the hydaulic elevator. A hydraulic elevator basic layout is shown on page 66. Apparently the device was designed during the early days of hydraulic mining although the first recorded instance I have been able to find of their use is in 1908 when a table of the efficiency of the elevators in use at the Wild Goose Mine near Ophir Creek in Alaska was published. The principle on which they worked was already being discussed in scientific journals much earlier than that so there is no doubt that hydraulic elevators were being used in the late 1880's when hydraulic mining was operating full blast.

The early development of the hydraulic elevator was brought about by the need in many mines to elevate the tailings to higher ground. As can be seen from the illustration, the lower sluice ran into the bottom of the hydraulic elevator and the tailings were elevated a considerable distance where they were run through another sluice. Although the height usually averaged around 25 feet, there are records of elevators lifting gravel and water 67 feet under heads of about 450 feet. As an example of how powerful a big head of water can be, Peele, in his classic work, "Mining Engineers Handbook," described an operation that elevated water and gravel 35 feet in an open sluice. So the rule that water doesn't run uphill can be disproven easily by a placer miner.

For those who like to get into such things, the first recorded writing that I have been able to find describing what resembles a modern dredge concerned a "gravel pump" which was used in Swaziland and Nigeria prior to 1915. This consisted of eight-inch centrifugal pumps which sucked both water and sand from a pit to a point 40 feet higher. The pumps were mounted on pontoons and the gravel ran through them. For those who are about to ask, the writer also mentioned that the impellers and the linings of the pumps were replaceable.

Although there are many cases of the hydraulic elevator being used in hydraulic mines well into the 1930's they really never developed into what we know as the modern portable dredge. This had to wait for reliable centrifugal pumps and dependable gasoline engines. Even when these were available, there was one more ingredient to bring this fresh idea to fruition. This was the rapid development of scuba diving.

By the mid-1950's it stood to reason that a scuba diving enthusiast who was also a mining expert had to come along. Such a person was the late Ernest Keene

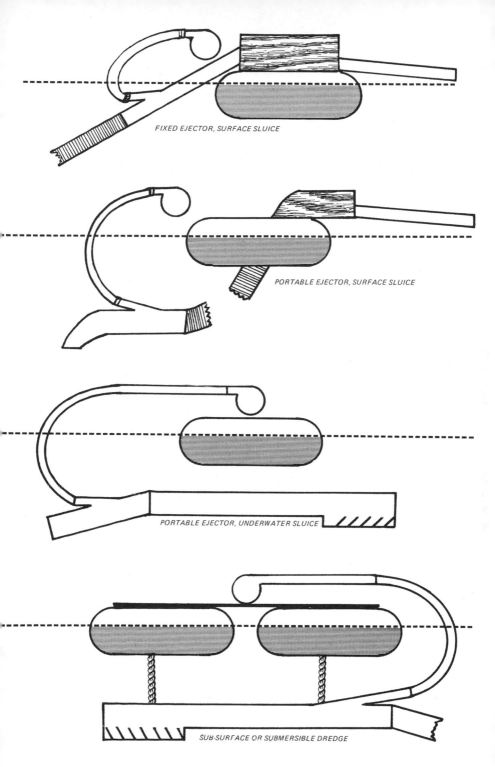

FIXED EJECTOR, SURFACE SLUICE

PORTABLE EJECTOR, SURFACE SLUICE

PORTABLE EJECTOR, UNDERWATER SLUICE

SUB-SURFACE OR SUBMERSIBLE DREDGE

63

who quickly realized that there had to be many rich placer pockets buried at depths which were beyond the capabilities of the placer miners of the past hundred years. He combined his knowledge of skin diving with his skills as a miner and his knowledge of hydraulics. It was the thought of supplying the water pressure by a high-powered centrifugal pump that really turned the trick. Within a few seasons in California's gold country, Keene had become a legend both for his exploits in recovering gold and his ability to design equipment which would do it faster and more efficiently.

Hobby gold dredging is far removed from the deep water operations of the mid 1950's. Today it is a family type of sport with thousands of individuals actively engaged, and since the price of gold skyrocketed, interest has increased remarkably. Many of the participants do it more for the thrill of finding gold than for the financial rewards, although the latter event occurs frequently enough to whet the appetite of anyone. Despite the fact that modern dredges are a far better piece of equipment; and is spite of the thousands of dredgers, the sport today is much the same as it was then. And, the debt modern dredgers owe Ernest Keene and the others of his kind is great.

Most persons who own a portable dredge are not quite sure just exactly how it works. They have their heads filled with terms like "Venturi effect" and "jet principle" and let it go at that. To cut though all this confusion in a few words, a modern portable dredge is simply a *jet pump*. The sluice, pressure box, suction hose, etc., are merely attachments to make it a combination sluice-dredge.

A jet pump, or ejector, as it is often called, is a pump that has no moving parts, which uses fluids in motion under controlled condtions to move another fluid or solid. Very simply stated a stream of fluid is forced through a nozzle to obtain the highest possible pressure. The stream is directed through a diffuser (called a mixer) creating a low-pressure area which causes the fluid being pumped to enter the mixer, join the high pressure stream and be ejected at some force. A commercial jet pump is illustrated in drawing 0-0. This type of pump is often used to pump caustic liquids, sludge, or any other material which might damage a regular pump.

The type of jet pump used in all dredges is called an *eductor*. This simply means a jet pump which uses a liquid as a motive fluid. An *injector* is a jet pump that uses a gas to entrain the fluid being moved.

Once the basic fundamentals of one type of portable dredge are understood, all others fall into place, even the hydraulic elevator. To explain these principles, I have picked one of the so-called jet-principle types where the exit orifice of the high pressure nozzle is positioned just below the water surface. Speaking from an engineering standpoint this configuration is just about the optimum for a general purpose surface dredge, and based on their popularity over the past few years, apparently gold dredgers also believe so. Through the next four chapters we will use this design to explain the fundamental operation of dredges.

A few notes of explanation are necessary before beginning. There are many forces which will not be considered in this explanation. These forces will be mentioned from time to time so that the dredger can be aware of them. Also, this description will, in the first part, consider only water moving through the dredge—the second part will explain how the gravel moves.

To orient yourself look closely at illustration 0-0. The various parts of the dredge are named in figure 0-0 and figure 0-0 contains the force arrows for the beginning of the description.

Force arrow C represents water being sucked from the stream where it is greatly accelerated in the centrifugal pump and forced at high velocity through the high pressure hose, force arrow D. Water from the high pressure hose, is exited through the orifice and proceeds up the ejector, force arrow E (This portion of the water is called the jet). During this process the jet pushes and sweeps along the water that was already in the ejector and represented as the difference between elevations I and J.

All of this water of the jet is discharged through the tube as shown by force arrows E and F. Note that there is an energy loss at this state of elevation but it is of little importance for the jet type of dredge as the pump is supplying considerable energy to spare. In fact, it must further dissipate some of this energy before the water goes to the sluice.

This surplus energy is largely dissipated when the jet strikes the opposite wall of the pressure box. Several things happen inside the pressure box. At the present only the force of the jet stream and entrained water is considered. In dissipating the energy, considerable turbulence is created and since the velocity is slowed somewhat and the volume of water now fills the pressure box almost completely, most of the air is forced out. (Air is still present inside the pressure box but its quantity is greatly reduced and most of it is represented by cavitation and water vapor at pressures lower than that of the atmosphere).

Another thing that you should remember is that not all of the energy of the jet is dissipated in the pressure box and water enters the sluice with some pressure behind it so that it now flows through the sluice faster than if acted on by gravity alone. This factor more or less seals the pressure box from outside air pressure.

Since the water being moved from the base of the jet through the pressure box and finally discharging to the sluice is, in effect, continuous with few cavities, we can now consider this portion of the dredge a closed tube, or pipe, with water flowing in at one end and flowing out the other. And, as will be shown later, water will be flowing continuously from the tip of the suction nozzle, we can consider the entire distance a closed tube, much the same as if a piece of pipe was laying in a stream with the current flowing through it.

Now I want to make a parenthetical note here so that many readers will not say to themselves—this isn't so. Of course it isn't, our imaginary tube of the dredge will be filled with pockets of air, water vapor, cross currents and turbulence in general. But the study of the movement of fluids is one of the most

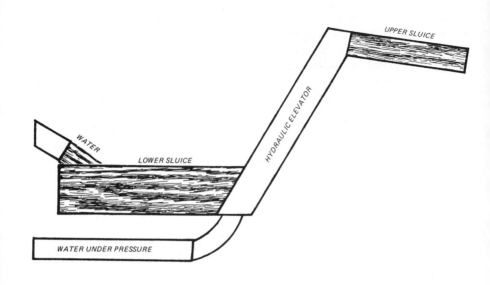

WATER

LOWER SLUICE

HYDRAULIC ELEVATOR

UPPER SLUICE

WATER UNDER PRESSURE

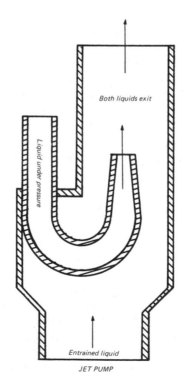

Both liquids exit

Liquid under pressure

Entrained liquid

JET PUMP

difficult of sciences and before scientists could really understand the dynamics of fluids they had to experiment and build their mathematical formulas to fit the results of their experiments. Even this didn't work too well, so they solved the problem in an unusual way. They simply invented an ideal fluid. An ideal fluid moves exactly as mathematics predict. It has no cavities, cross flows or turbulence. Once the scientists had the ideal fluid, they translated its movements into a "real fluid" which is the way a fluid, or in this case, water, actually behaves.

It is in this context that the explanation of how water moves from the intake is being discussed—I'll make some notes on the differences of the real fluid flow later in this chapter.

Now that this small item is taken care of let's go back to our imaginary point A to G in illustration 0-0. If we lay it in the water at an angle of 60 degrees as the picture is drawn, leaving about two feet sticking out of the water, after a few seconds the water level inside the tube will be equal to the water level outside of the tube.

Now imagine that you have a plunger that can be inserted 24 inches below the water level and tightened so that no water can escape when it is removed. In other words, when you quickly pull the plunger back up, it will remove all the water in the tube above the plunger. But look again at the tube, as fast as you removed the 24 inches of water, another quantity of water appeared to fill up the empty space created by the quantity you removed. In fact, the tube would fill up almost as soon as you removed the water.

This filling up of the tube is because the pressure of the water was greater at the bottom of the tube than it was inside the tube. And, even if the tube had been inserted only about 25 inches, the water level would have returned back to normal. Water seeks its own level and will tolerate absolutely no difference in levels unless there is some difference in pressure. Equalize the difference in pressure and the water level equalizes very quickly.

If you want to prove this principle in reverse, get one of the plastic "see-through" straws that are sold in super markets. Fill a glass with water and hold your thumb tightly over one end and insert the other in the water. Keep the straw perpendicular so that no air can escape from the bottom end of the straw. As you push the straw deeper in the glass, the water level in the straw will drop below the level of the water in the glass. Raise the straw and the water level will come up. What happens, of course, is that the air pressure in the straw is being increased as water comes in at the bottom level and starts exerting a force greater than air pressure on the water inside the straw. As more water enters the straw, the pressure becomes greater and the level is even lower. Take your thumb off the straw, the air pressures become equal and the water inside the straw becomes level with that of the water glass.

By now I am sure I have given away the great secret and almost all readers are beginning to understand how a dredge works. But please bear with me, there are many more important facts.

Now that I have explained that we can consider the dredge the same as a water pipe, I am going one step further and say let's consider the water in the dredge suction tube as a column of water. Under these circumstances we can calculate the forces inside and outside the pipe quite simply.

Let's return to point E in the ejector. Since the water being moved from this point has a force greater than that of its own gravity, the gravity of the water already in this portion of the tube, and the pressure of the air pushing down on the mouth of the tube, we can safely assume that at the base of the jet there is no weight—or to be more precise, there is no pressure or force acting downward on the remaining water in the tube.

This means that the only downward force operating at the exact point of the suction nozzle is created by the weight of water inside the tube.

For all practical purposes our hypothetical dredge has an inside diameter of four inches and a length of 12 feet. Taking the water below the jet and including the triangle of weight shown in the illustration, we would find that this water exerts a pressure of about 4.4 pounds per square inch. Taking any column of water outside the tube and adding the pressure exerted by the air, 14.7 psi, we discover that this column of water will exert a pressure of 17.3 psi at a depth of six feet which is exactly where our 12 foot intake hose is going to be if it is at an angle of 60 degrees with the stream bottom.

There is little need to explain that the greater pressure will force water into the tube very quickly. In fact, it will be difficult to remove the water from point ?? fast enough to keep water from constantly maintaining a level with the stream level.

The careful reader will immediately detect that I have only described exactly how the water level is maintained inside our pipe with its column of water. And, it is well to remember that regardless of the differences between the column of water in the dredge tube and a similar quantity of water next to it, these differences can only do one thing—provide a constant supply of water to be moved by the jet. But this is enough.

Remember the point I stressed early in this chapter? That the jet of water picked up and evacuated all the water from the area B of the tube. Now that we know that this area is going to be supplied continuously with water from the pressures at the suction nozzle head, we now realize that the jet is going to be evacuating this water continuously to be replaced with the water from the bottom of the stream.

Actually what we have created is a stream flowing at 60 degrees to the normal flow of the stream and we have made it flow uphill. If you will return to the example given earlier of the plunger pulling out water from the top 24 inches of our imaginary pipe. Straighten out the pipe so that it is perpendicular with the stream, put a check valve in the piston, and hook the piston to a handle.

And, you have an old fashioned pump—not too far removed from our modern jet-principle dredge.

The process by which a jet pump moves water is called entrainment, which simply means to catch up and carry away. How the jet entrains the surrounding water is not fully understood but there are four principles which are involved. Only three of these are considered for water. They are: 1. The acceleration of the particles of the stationary fluid through which the jet passes; 2. Capture of the stationary fluid by viscous friction on the outside of the jet; and 3. The rapid expansion of the jet to a pressure below that of the stationary fluid. The fourth cause is found only in gases such as steam, and involves chemical changes due to different temperatures.

Although the explanation is quite lengthy, the whole story is quite simple. The jet pump evacuates water out of the top portion of the suction tube and it is constantly replaced by pressures pushing more water in the intake nozzle thereby providing a steady supply of fresh water to be pumped out.

Naturally once the column of water in the suction tube begins moving, the laws of conservation of energy and the principles of momentum and impulse must be satisfied. Once the water is moving the stream is constant from the suction nozzle to the point of discharge through the sluice.

A study of the mechanics of the jet pump is beyond the scope of this book and most of the facts are only of interest to the engineers of the dredge. Actually the jet pump is very inefficient which is why there are so few of them used in industry. Some of the tests made on jet pumps show them to vary from two to 25 percent in efficiency compared to conventional pumps. The best results were obtained when the head was operating at 90 percent of efficiency.

Would the jet pump work if it was placed above the water level? The answer is yes and some modern dredges are designed this way. Its efficiency is lowered somewhat because it has to pull an additional quantity of water in proportion to the water it pushes to the pressure box. Since most dredges have a surplus of energy, they seem to work well in the smaller intake sizes. The major advantage of this configuration is that it can operate in shallower depths than when the ejector is located below water.

I have tried to track down the theory that the dredge works on the venturi principle in many sources, but to no avail. Apparently the wide belief stems from the fact that jet pumps are frequently used to mix liquids or gases. In this case a ventrui tube (actually a smaller diameter tube inserted in a pipe) is installed just after the nozzle of the jet. When the two liquids or gases enter the venturi tube they are rapidly mixed due to a specially designed taper. There is no venturi tube in a modern dredge.

It may have occurred to the reader about now that we have only explained one type of dredge operation—the so called jet dredge. There are others, of course, and they go by many names, underwater, suction, etc. Let's take the two major types and look at them before moving on to the transportation of material.

The first of the gold dredges were the underwater sluice type. Today many of them are called submersible but they have changed little over the years. In this type of dredge the centrifugal pump remains on the surface and water is directed through a long high pressure hose to an underwater tube which picks up material at its suction nozzle and moves it through the tube and over a small sluice which is an integral part of the suction tube.

The underwater dredge is also a jet pump and the suction developed is often equal or even better than that of the ejector mounted higher type for the same head supplied to the jet pump. Since the jet does not have to overcome air pressure and gravity due to operating in a level position, all of its energy can be used to push the water in the tube and entrain water at the base of the jet. Strangely enough, air pressure is of no advantage to the underwater dredge because its effect works on both the intake and discharge side of the suction tube. Neither is the flow of the stream of much help to the head of the jet for the jet must dissipate its excess energy into the stream. A few feet beyond the discharge exit it slows to the velocity of the stream. This creates a pressure inside the tube which must be overcome.

This is not a disadvantage, however, the jet supplies more than enough energy for the tube and the suction created is powerful. Experiments have shown that the same jet head supplied to a three inch, near the surface ejector is sufficient to power an underwater dredge with an intake capacity of five inches. This means that the underwater dredge will move more gravel quantity for a given time than the other types with the same velocity jet.

One of the disadvantages of the underwater jet is that due to its design the sluice must be smaller and hence, less efficient. The dredger is moving more material over a smaller sluice at greater speeds and this means a loss of values, especially the fine values.

Due to this peculiarity, manufacturers have developed the combination dredge. This is a fixed, near the surface ejector, surface sluice type of dredge basically. As an accessory the dredge is equipped with a longer high pressure hose which attaches to the underwater suction tube.

In the latter configuration the dredge can be used to move a considerable amount of overburden in a shorter time. When the dredger decides he is near the area where higher values may be found, he merely changes high pressure hoses and switches to the surface sluice.

There are other disadvantages to the underwater dredge. One of these is that it is unwieldly and hard to negotiate around or under boulders. Also, since the point of discharge is much nearer the operator he often finds he has to re-dredge some of the tailings if the area of overburden is extensive. However, neither of these shortcomings are great when compared to the shorter time to remove a large amount of overburden and many larger operations now use the combination dredge for bigger operations.

The underwater dredge has theoretical unlimited depth and at times is the only type of dredge that will work for a scuba diver. The centrifugal pump should be able to pump water to any depth as long as the pressure of the water does not collapse the high pressure hose or suction tube. This would be at a depth greater than a diver could work so from all practical purposes, the underwater type sluice can be considered efficient at any depths. Many cases are known where considerable gold was recovered from depths of sixty feet or more.

One of the advantages of owning a fixed ejector sluice is that at a later date, when time demands, the underwater sluice and longer high pressure hose can be added with little cost.

The first development from the underwater dredge was to connect the ejector to a surface sluice. This is the closest that modern dredges come to the old time hydraulic elevator and the only basic difference is the movable ejector.

In this configuration the jet enters the tube just above the head of the suction nozzle and the jet has to move the entire column of water to start the momentum of the stream. This consumes energy far quicker than does the higher mounted ejector simply because there is now more entrained water for the jet to push. One major advantage for this type of operation is that the weight of the water inside the tube from the base of the jet to the mouth of the nozzle is much less causing pressures outside the tube to force water in much faster. This advantage is good, but not enough to overcome energy losses higher in the tube. For an equal length of suction hose the higher mounted, fixed ejector is more efficient.

To understand why this is so we must remember that the energy of the outside water which pushes water into the tube is potential, or stored, and cannot be used until the pressures inside the tube are less. So this means that before all this energy can be used, and add to the momentum of our column of water inside the tube, water must be pushed out by the jet. This is the reason— simply that the energy of the jet can evacuate the shorter column of water in the mixer more times per minute than it can the ten foot column of the suction hose.

This preceding principle has often been explained by stating that the suction is gained through less friction in the various pipes. This reasoning doesn't stand up under investigation. The frictional forces in the suction tube are the same regardless of where the jet nozzle is placed. Since nearly the same quantity of water passes through in each situation these losses or gains must equal zero. The high pressure tube probably has a coefficient of friction equal to or less than that of glass, which is about .02. This means that if the pump is discharging 100 gpm at its outlet, the discharge through a plastic pipe about 100 feet long would be 98 gpm. Since the length of the movable ejector high pressure tube is seldom more than 18 feet long, and considering the fact that the water inside is moving down so that gravity also accelerates it, it is doubtful if there is any loss of head at all.

Actually you should not be too concerned about all these pressure losses I have described. Modern dredges are engineered with considerable energy to spare for most situations and it is only when they are being used to maximum capacity that the differences in efficiency become significant.

If the efficiency of the movable ejector is less then why hasn't it disappeared from the scene? Well, there is a simple answer. No other type of suction nozzle will work as well in shallow water. Many cases are known of this type of ejector working well in less than six inches of water. All that is required is enough depth to cover the mouth of the suction nozzle and enough water to supply the suction pump.

Now that you understand the way different types of dredges work it is time to explain just how they "suck" up gravel and sand from the stream bed. Actually nothing is really sucked up, it is entrained just as is the water near the jet of the jet pump. The gravel is suspended, pulled into the suction nozzle stream where it is swept away by the momentum of the current.

Those readers who have digested the section on stream transportation already know that when a current is flowing it will begin to move the sands of the bed, first by saltation, and then the velocity will suspend the sands and transport them downstream as though they were floating. This same thing happens near the edges of the suction nozzle of the dredge. Outside pressures push the water toward the nozzle with its lower pressure thereby creating a small current similar to an eddy. This raises the sands where they are suspended and swept away in the current of the suction tube.

Quite large stones can be moved in this process. Usually what happens is that the sand around the stone is removed and the pressure from the current begins concentrating its force on the stone. When there is enough force beneath the stone it is suspended and swept up the tube. A well engineered dredge will pick up a stone more than double its suction nozzle at times, often with bad results. This subject will be dealt with more in the chapter on how to operate dredges.

Why does gravel move up the suction tube with such great ease? The answer is quite simple. First of all newcomers to sluicing think of the weight of a stone as it is in the air. Underwater this weight becomes greatly reduced due to the principle of buoyancy. This reduced wieght is a considerable advantage in getting the stone or gravel to the jet where it is accelerated even further and easily gains enough momentum itself to travel to the pressure box.

If the reduced weight were the only advantage the dredge got, it would not work quite as well. But remember also the law of the inclined plane. This effect does not reduce the weight of the stone but reduces the forces required to keep it moving once acceleration has equalled the velocity of the moving column of water.

Let's see how it works. Suppose the dredge manufacturer has designed his four inch nozzle so that it will only take a three inch stone in order to reduce clogging in the tube. This means that the maximum stone that could enter the nozzle could be in the shape of a perfect sphere. Such a stone of granite would weigh 3.7 pounds above water, 2.3 pounds below (Give or take a few tenths depending on the density of the granite and the temperature of the water). But to move it up a suction tube inclined 60 degrees with the surface of the stream would require only about two pounds of force to keep it moving at the velocity of the water.

In explaining how a dredge works I have omitted many things like cavitation, vapor pressure, turbulence, cross currents, etc. All of these things are happening inside a suction tube and pressure box but in understanding how a dredge works they become insignificant compared to the major forces. Although these phenomena can be ignored for this analysis they serve a very important function.

Anyone who has operated an ordinary sluice knows that often the gold bearing material is lightly cemented, sometimes clayey, sometimes harder. During a hand operation this material must be broken up and often run through a grizzly. All this is taken care of inside the tube where turbulence and cross currents start the job at the base of the jet. Here sand and cemented materials are broken up by the force of the water and larger pieces of gravel continue the process during the trip up the tube as they crash into each other. When the force of the jet is dissipated in the pressure box under tremendous agitation due to turbulence, usually even firmly cemented material is broken up.

At this point the first classification usually begins. Most dredges are equipped with a small grizzly or sorting screen which allows most of the smaller particles to drop through to the floor while bigger rocks rush out. Since the sand which usually contains the gold particles is directed to the lower portion of the sluice stream, the gold is more easily separated and retained.

Before moving on to setting up and testing a new dredge this is a good place to describe the four basic types of dredges currently available so that you will have all the information in one place to help in making a decision on which type to buy. In making such a decision you should remember that once the float, pump/engine combination and sluice are decided on you can then make several decisions on what type of a suction nozzle. For instance, if you start with a fixed ejector, surface sluice, it is possible to later add a portable ejector for shallow operations—or to add an underwater dredge for removing major overburden.

If your plan is to add different suction nozzles at a later date, discuss it with the manufacturer before you decide on the basic dredge. The capacity of the pump and sluice is the key, and many manufacturers are now offering combination dredges as a unit, so matching the pump and sluice to future suction nozzles should present no problem. If you must purchase a pump that is rated higher than the capacity of your initial suction nozzle, it can usually be slowed down by reducing the engine speed.

FIXED EJECTOR, SURFACE SLUICE. This type of dredge is the most popular at the present time. They are lightweight, efficient and reasonably priced. Most are engineered so the ejector removes quickly from the pressure box so that later modification to a different suction nozzle can be made. Some have their ejectors just above the water level but the most popular configuration is to have the base of the jet about a foot or more below the surface. They are available in several sizes and pump capacities. At the lower end of the range are the 1½- and two-inch models. Some of these are light enough to be toted to remote locations by one or two individuals backpacking. Many experienced and sophisticated gold dredgers use them as prospecting tools, and dredge areas along the stream marking those that are worth bringing in heavier equipment.

The middle range of the dredges are those in the 2½-, three- and four-inch sizes. The selection usually depends on a person's checkbook, the size of the operations he intends to conduct and the amount of space he has for transportation. The smaller ones can usually be put into the trunk of the average sized automobile and the bigger ones in a normal station wagon. Floats are always a problem. Plastic ones can usually be transported by use of a roof rack; the inner tube type can save considerable space by being deflated for transporting.

At the professional end of the range are the five-, six- and eight-inch versions. (Sizes over eight-inch are available, but usually on special order only.) These are strictly for the fellow who is making a career out of placering, or the individual who has a known deposit he wants to clean out in short order. Some of the eight-inch versions have two centrifugal pumps putting two jets into the ejector. The suction on these is quite powerful and many of them have been used in depths of 50 feet or more.

PORTABLE EJECTOR, SURFACE SLUICE. At one time the only alternative to the underwater sluice, these dredges have taken second place to the new fixed ejector, surface sluice type. While not sold in the same quantity as before, they have survived as accessory suction nozzles for this type of dredge, especially in the three- and four-inch sizes which operate well in shallow water.

PORTABLE EJECTOR, UNDERWATER SLUICE. These are the granddaddys of the dredging pastime. They are still popular for moving a lot of overburden or for deep water diving operations.

SUB-SURFACE, OR SUBMERSIBLE DREDGE. This is a true combination type dredge which is generally engineered to work for an underwater operation only. The suction tube and sluice is suspended from a float which also mounts the centrifugal pump. It is useful in deep operations since the material has to be transported only about half way up. It is easier to operate over long periods as only the suction hose must be manuevered and not the entire sluice box.

10

SETTING UP AND TESTING YOUR DREDGE

The internal workings of any modern dredge are a complex procedure but getting the water flowing is really very simple. Almost all the physical processes described in the preceding chapter start almost instantaneously a few seconds after the engine is turned on. But only by understanding each element of the operation can a person eliminate the many problems that inexperienced dredgers have in learning to operate their dredge.

It is not a good practice to do the first assembly or testing of a dredge in the field. Usually a dredger's home is many miles from the spot he wants to dredge and if anything goes wrong, it could lead to an unsuccessful trip. This chapter will describe how to set up the dredge and test it but please read it through before unpacking your dredge. Once you have your new dredge launched and start the engine, things will happen so fast there is little time to analyze anything.

The first thing a new owner must do is to read all the manuals the manufacturer has supplied. Traditionally these manuals cover the complete assembly of the dredge from the first piece bolted together to the second and so on until the last nut is tightened.

At first you might get the idea that your new dredge is a kit, and this is not a bad analogy. There are a couple of good reasons why it comes broken down into so many pieces. First, the shipping is much less and the danger of breaking fragile parts by a common carrier is greatly reduced. Second, the manufacturer has no way of knowing how you will transport your dredge. Some may have to stow it compactly in the trunk of an automobile while others might be able to transport the dredge as one unit in the back of a pickup truck. You will probably fall somewhere in between.

After you have studied the manual, the first step is to assemble the dredge in your backyard or some other convenient area. After the dredge is put together analyze the completed project and the processes which went into assembly according to your own needs for transporting. This means that you may be able to separate the dredge into a half dozen sub-assemblies which require only eight or ten fasteners, instead of over a dozen requiring many individual bolt and tighten operations. This is something each individual must do for himself as no one can predict another person's needs. The fewer parts to be assembled, the quicker the dredge can be put into operation.

Once you have decided how many components you want to break it down into, the next step is to test it so you can find out if everything is working.

Ideally this initial launch should be done in still water which is at least deep enough to clear the intake hose by 12 inches or more. The best solution is to have, or find a friend who has, a swimming pool. If none is available the next best spot is a local pond or the slow moving portion of a nearby stream. It is important at this stage of the game that the pond not be too deep—about waist deep is ideal—so that it is easy to work and move around during the familiarization aspect of the procedure.

If you have to try your test in a moving stream be sure to read the next chapter on anchoring before launching.

Once the dredge is successfully launched it is time to start the engine. But before we can do that in this text we must explain what is going to happen. The main reason we wait until the launch time to start the engine is that the centrifugal pump must never run without a constant supply of water. Actually the centrifugal pump will run for a few moments without water but the time is very short. It needs water for lubrication and cooling of the impellers. If run even for periods no longer than 30 to 60 seconds, it will heat rapidly and may cause permanent damage.

No centrifugal pump is self-priming, even if its intake is within a few inches of the water. The pump can be primed by the simple expedient of letting water enter through the high pressure hose. But this means that the high pressure hose must be disconnected at least in one place and then reconnected. The idea is to get going without having to tear the dredge down several times.

On virtually all dredges the pump intake hose is equipped with a foot valve and strainer, illustration 0-0. This is usually a butterfly type valve which opens when there is pressure on the bottom side and closes when there is no pressure. The tension on the valve is slight—just enough to close it.

To prime the centrifugal pump, move the end of the pump intake hose rapidly up and down under the surface of the water. On each down stroke the foot valve opens allowing water to flow into the pump intake hose. On each up stroke, the water is trapped inside the hose by the closing of the butterfly valve. The water trapped inside the hose has a small momentum and tends to stay where it is during the next down stroke creating a small low pressure area which speeds the intake of water as the butterfly valve opens. Moving up and down like this the foot valve is actually a small piston pump which forces water up the intake hose quite efficiently. To prime the centrifugal pump it has to elevate the water only a foot or two and this requires little effort.

Once the priming water enters the pump, the impellers thrust it outward towards the pump housing at great speed creating another low pressure area which pulls water inside the intake hose to the impellers with greater momentum. This causes the foot valve to remain open and water flows constantly to the impellers. All this takes some time to explain but in practice the whole operation takes only a few seconds of rapid up and down thrusts.

The mistake most beginners make is not making the up and down thrusts rapid enough so the momentum of the water inside the intake tube can do its work. Agitate the foot valve rapidly enough and the surge of water through the pump will occur so quickly that it will surprise you. The biggest enemy of

76

priming the pump is air which is contained inside the pump and high pressure hose. This air must be evacuated and is generally compressed. I'll cover this problem later in this chapter but in most cases, priming the pump is as easily done as described here.

It may seem strange to the reader that I have explained the operation of priming the pump before discussing the engine. There is good reason for this. It has been my experience that the beginning dredge operator experiences his first problems in priming the centrifugal pump. But it is essential for preservation of the pump that it be primed quickly. Most newcomers start their engine and then try to learn how to prime—sometimes with disastrous results. After launching your dredge for the first time, try priming the pump in the manner described before starting the engine. Remember, you may have gotten the one foot valve in a thousand that is defective or installed wrong.

Now that you have launched your dredge and have an understanding of how the centrifugal pump is primed it is time to start the engine and get all of the physical movements described in the previous chapter working—in other words to activate the system.

I cannot cover the operation of the gasoline engines commonly used on dredges in this book. There are too many different models and manufacturers. Often the requirements change from maker to maker and even from model to model of the same manufacturer. So only general rules can be given.

If your dredge is new, the engine should have a separate owner's manual which should be contained with the literature supplied with your dredge. If there is none supplied—insist that the dredge manufacturer get you one.

Study this manual carefully and follow as closely as possible all the manufacturer's recommendations. The manual will give the rules of maintenance that are often ignored by the fellow who has his engine attached to a lawn mower. Your engine is probably going to be a little more high powered, more sophisticated and, most important, used for long periods of continuous operation. Following the manufacturers recommendations for oil change, filter cleaning, etc., can mean many more hours of trouble free use.

One important thing to remember is that all new engines are delivered without fluids. This means no gas and no oil. Manufacturers usually spell this out in their manuals in bold face type—but it must be understood by all new dredge owners. Also, there is no guarantee that the last fellow who owned a used dredge left the proper type or quantity of oil in the crankcase, so check this out before starting a used engine, too.

Most modern engines used on dredges require no break in—or very little—and this can usually be done under the load of a pump. Don't rely on what the engine manufacturer tells you, however. Ask your dredge manufacturer—if there have been any problems due to improper break in, he has probably had to solve them for his previous customers.

Once the engine is operating and the pump primed, things happen very quickly. As soon as the centrifugal pump begins discharging the jet is established and the jet pump begins to pull water through the suction tube. Some jet pumps must be primed but before discussing that let's cover another point which could be called the final check on establishing the jet.

The first inspection is to examine the discharge of the pressure box through the sluice. If everything is working right it should fill the sluice about two-thirds of the way and be moving at a respectable rate. (This is not true if the suction tube must be primed and only the jet is flowing through the pressure box, so this must be done before gauging the quantity of water through the sluice.) If water isn't flowing in quantity and at speed, or if it quits almost immediately, the pump must be primed again or the engine shut down for inspection. At this point there is no time to look for anything else, get the water running by re-priming or shut down the engine.

If the water is running through the sluice at a fair rate, it can be safely assumed that the pump is getting enough water to lubricate and will not be damaged. Then you can proceed to the next inspection—that of the high pressure hose.

Just before we primed the pump, the intake hose, pump and high pressure hose contained air. Air, or any gas, is the enemy of a centrifugal pump designed to pump a liquid. Therefore before your centrifugal pump can begin to produce the results the manufacturer promised you, all the air must be evacuated. Many manufacturers and writers recommend loosening the clamps of the high pressure hose to allow air to escape. I do not recommend this for the following reasons.

First, when the engine is initially started the impellers force most of the air out into the high pressure tube where it is slightly compressed. Air from the intake hose is also pulled out and contributes slightly to raising the water inside the intake hose. The quantity of this air is quite small, only a fraction of a cubic foot. The inrushing water quickly entrains all the air and evacuates it with the initial surge of water through the high pressure hose.

Another reason is that once the pump starts operating it reaches its maximum capacity very rapidly and the pressure is extreme—that's why they call it a high pressure hose. If the hose is loosened at the pump, the pressure can easily make it slip and come off and it is practically impossible to re-install it. If the hose is loosened at the ejector, the pressure tends to straighten the hose out and bending it back to reconnect it is quite an effort. In any case, if it comes loose you are going to get soaking wet.

When the pump is naturally evacuating the air from its system the results are shown in the clear plastic hose as turbulence, usually in the form of small bubbles. There is little possibility of the entrained air forming as a consolidated mass of compressed air resisting the flow of water. If this did occur the water surrounding the mass would be at vapor pressure and invade the air pocket causing cavitation. If cavitation occurs in the high pressure hose the large mass of compressed air would alternately collapse and reform as one or more small masses so many times per minute you could not see it happen. What you would be able to see would be small bubbles moving down the current flow with considerable rapidity. This is not unusual inside the high pressure hose when the flow is just starting.

So during the initial inspection of the high pressure hose, look for turbulence and rapidly moving bubbles. These should disappear almost completely in the first ten or twenty seconds of operations.

Assuming a clear plastic material so it can be inspected visually, when the high pressure hose is operating properly it should appear as full of water but give little indication of movement. When squeezed between the fingers it should offer great resistance and not buckle. If there is a lack of resistance, or if bubbles persist (in quantity) the centrifugal pump is probably sucking gases of some kind.

To cure this problem, first check the clamp of the intake hose to make sure it is tight. If there is any doubt in your mind, dip your hand in the water and dribble water on the clamp area. If it sucks in the water, the problem is that the pump is sucking in air at this point. Tightening the clamp should solve it.

If this doesn't work, run your wet hand over the portion of the suction tube above the surface of the water to see if there is a leak here. (Note: In this and subsequent lead tests, oil works better as it is easier to see what is happening. However, it is quite messy and most dredgers use the water test.)

If both of these tests prove negative examine the position of the foot valve and strainer. It may be that the strainer has moved to a place where there is underwater turbulence and cavitation that is not visible to the eye from the surface. If this is the case, vaporized water may be being sucked into the pump where it can not only slow down the pump velocity, but damage the impellers and housing as well. To cure this problem simply move the strainer to another, more quiet portion of the stream.

If these three tests fail, it may be that you have a defective pump housing. If this is so there is one test that can be made in the field with little difficulty. First examine the outer circumference of the pump housing. Any cracks or minute casting imperfections here will show up as bubbles or small jets of water being expelled through the housing. Since no air would enter the housing due to such a defect it is improbable that minor leaks here would cause much pressure drop so it is unlikely that a loss of water velocity would be revealed here.

The final test is to drop water around the intake inlet junction where it joins the pump housing. You should be able to see the water being sucked into the pump which would indicate a defective cast.

Naturally if any of these simple tests reveal a defect in the pump or suction hose the bad part should be returned to the manufacturer for replacement.

However, it might happen that all of these tests are negative and all components check out in the field. If this happens it is best to shut the dredge down and rest a moment—perhaps moving to a new location. Then start up again, carefully priming the pump and doing everything text book correct according to the manufacturers instructions.

If the quantity and velocity of the sluice is still low, or if a large quantity of air still appears in the high pressure hose, it is reasonable to assume that something inside the pump housing is defective. It may be that the impeller casting is bad, the bushings out of line—or any of several other things that could cause the impeller to rotate unevenly causing turbulence and cavitation. If this is the case, further inspection by the average dredge owner is outside his realm of knowledge and the pump should be returned to the manufacturer with a full explanation of what is happening so the maker can remedy the situation.

In the case of an extremely long suction or high pressure hose where the quantity of air is considerably more, it is a good idea to eliminate as much of the air as possible by filling the hoses with water before turning on the engine. This is especially true of the movable ejector type which often has a pressure hose 15 feet or longer. As a last resort to difficult priming situations, the entire system can be filled with water before starting the engine. This leaves only a minute quantity of air inside the pump to be evacuated.

It is regrettable that each operation in a dredge must be explained in such great depth before another procedure is explained. In practice it takes only a few moments to get a dredge system going. In describing the next step of an instant it is often necessary to separate the operations by considerable verbal distance.

Technically by priming the pump all we have done is to establish the jet. To get the entire system going it is necessary to also prime the jet pump.

There are two basic positions for the ejector: above water and below water. All below water types are self priming and once the jet has been established, the system is operative. The above water jet, however, needs to be primed and there have been many lengthy explanations of how this is done. One suggestion is to slow the engine down and hold your hand over the jet discharge, forcing the water down into the air space. This works but anyone who has tried to hold back a stream discharging even 50 gpm realizes just how difficult it can be— besides usually soaking you. In the open sluice type with no pressure box it might make sense, but with the pressure box system (which must be hinged to be practical) it is very hard. Once the system is operative there is the problem of closing the lid.

Fortunately there is a simpler way. First establish the jet, making sure the suction hose is entirely underwater. Then grasp the center of the flexible suction hose making certain the suction nozzle is kept below water at all times. Lift the center section of the suction tube a few inches above the height of the base of the jet. If you do this with a gentle action you will establish a siphon effect and the water will raise to the suction area of the jet pump with no cavitation. When this happens the air and water in the tube will be quickly entrained and evacuated. Water rushes into this low pressure area and the system is operational.

It is doubtful if any but a small fraction of new dredge owners will ever have to perform all the tests given above. In fact, I have yet to see a dredge operation that had to do little but check the high pressure hose for turbulence. But when problems do arise, the tests given will usually locate the trouble.

I have taken several pages of text and illustration to explain one thing. How to get water running through the complete system of a dredge. By now a newcomer who has not yet purchased his dredge might assume that it is really a complex thing—hard to learn and hard to do. To dispell these doubts let me summarize just what is involved.

1. Start engine.
2. Prime centrifugal pump.
3. Examine high pressure hose, jet and sluice.
4. Prime jet pump if necessary.
5. Adjust engine speed for gravel and depth being worked.

It is actually that simple and once the engine is started the whole process takes only a few seconds. Note that I have added one more procedure, Number 5, to the list that we have not yet considered. I mentioned this step here so you would be aware of every step necessary before actually starting to dredge.

Adjusting the engine speed is a function of the gravel being worked, the distance it is being transported to the sluice, and the altitude in which the engine is being used. There is no way to safely predict how fast you will have to run your engine until you are actually in a dredging operation. At this stage of the familiarization process, though, you may want to experiment a little further to see what differences the speed will make. Remember that everything will work a little better now since we are talking about just running water and not gravel through the dredge. To speed up the experiment, have a friend change the speed and show him where the kill button of the engine is located.

Try several different speeds of the engine. First put your fingers over the mouth of the suction nozzle and feel the power of the suction. Do this carefully as at full speed the suction can trap your hand on a small tube—or even pull it into a larger one, resulting in injury. If this happens, have your friend hit the kill button.

It takes little experimentation to realize that your dredge has pretty strong suction even at slow speeds. If you want further proof get a stone smaller than the size of your suction nozzle (Don't use a pail of sand. The dredge discharges back into the pool and if it is a swimming pool, it's much easier to find a rock than a shovel full of sand scattered over the bottom.) Simply release the rock in the suction tube and see how quickly it reappears in the sluice.

This simple experiment should give you confidence that your dredge has suction power. But before quitting on your speed tests move your attention to the sluice. At full speed, in water with no gravel being transported the water should almost fill the sluice. In some cases the water may even overfill the sluice.

Now slow down the engine until the water in the sluice starts to drop. To notice much reduction under these test circumstances, you will probably have to slow the engine to less than half speed. At a point when the water reaches about two-thirds of the depth of the sluice you will be at the least practical speed you can safely use. Don't worry if your engine stalls and quits before you reach the two-thirds level. It merely means that the operating range you have will be more than enough power to move gravel through a wide range of engine speeds.

Of course, if your engine only produces enough flow in a water-only situation to fill the sluice box half full at peak rpm, there is something wrong. Usually this is a matter of carburetion or other minor adjustments. If this is the case have your engine checked by a competent mechanic. If your engine isn't working at peak performance in your back yard, you can hardly expect it do well in the field. This is especially true if you are going to be dredging at 5,000 feet or more where power losses due to altitude can be expected.

By now, if you have followed the steps outlined in this chapter, you will be quite familiar with the way your dredge operates and how to set it up and get it running without too much difficulty. The next step is to go to a gold bearing stream and start dredging.

11

THE INITIAL LAUNCH

Describing the initial launch of the dredge at a gold bearing location is one of the most difficult procedures to put into plain text. Each launch is different and each almost pre-determined by the natural layout of the stream you choose to dredge. There are a few general rules, however, and these will help on any type of launch.

A successful dredging operation begins at home in planning the trip. The selection of camping equipment and supplies is not part of dredging, but the choice determines a good deal how much other paraphernalia can be taken so the only advice that can be given is to choose wisely.

Most dredgers I know go light on the camping luxuries simply because they want to pack as many tools as possible. Among the supplies that will be needed for the actual dredging are gas and oil. Personally, I do not transport the gas a long distance, taking only empty cans which I have filled near the dredging location, unless, of course, there is a gas shortage. Oil is another matter. To avoid leaking and casual spills inside my vehicle, I most often drain the oil from the engine and refill it at the dredging location. This might sound unusual but it has sound reason. Oil splashes around and can get into the piston and carburetion area of the engine during loading and unloading, creating hard starting and a necessary warm up before the engine is operating. Refilling at the dredging location avoids this problem. If I am traveling in a pickup truck, however, where the engine can be loaded easily and transported in its normal position, I feel that little oil will thrash around and leave it filled.

While the quality of gasoline is usually reliable even in small towns, many stations will not carry the particular brand of oil I want. Whether the engine is to be filled or not, I always take enough oil for several complete oil changes in case of an accidental spill, or for long runs.

Besides fuel, the second consideration for a dredging operation is a good supply of tools with which to work on the dredge and engine. As a bare minimum, pliers, a couple of screwdrivers and a crescent wrench will perform most jobs when combined with the normal tools of placering. In my case, I purchased a bare minimum automotive tool kit from a hardware store which includes such items as hacksaw, socket wrench sets, alan wrenches, etc. With this assortment I can completely tear down the engine if necessary. Although I have never used many of the tools, the confidence having the set gives me in case of breakdown is well worth the cost.

In addition to the tools, give careful consideration to a supply of nuts, bolts, washers, wire, short bits of chain, and any other item which might seem to you to be useful when you are forty miles from a hardware store. This is a point well worth considering carefully. Most dredging locations are far removed from stores with a sophisticated stock of mechanical supplies and it seems to be a rule of shopping in the back country that the store you find has everything but the item you need worst.

I try to have an extra supply of each type of nut and bolt that the dredge has. After some experience in dredging, I also accumulated almost every nut and bolt that the engine has. Beyond this, the choice of items is pretty much up to the individual dredger and you will eventually add to your collection the items that you needed and didn't have on a particular trip. Here are a few goodies that have come in very handy for me. First is a small roll of fairly flexible wire and one of medium size. Always take plenty for no other item comes in so handy in jury rigging. I also carry two or three pieces of chain varying in length from four to ten feet with heavy bolts and nuts to attach them to the dredge.

Many repairs on the float and sluice can not be made with a conventional nut, bolt and washer. In the automotive section of most hardware stores they sell an oversize washer called a fender washer. It is used to bolt sheet metal to other items and will often be the only way repairs can be made on thin material. Another hardware item that is very useful is the eyebolt. This device is shown in figure 0-0. The bolt portion has a long thread and by using two nuts, can be bolted onto many different areas of the dredge to provide anchor locations either permanent or letting the rope or cable slide through. In picking eyebolts be sure that the eye portion is about twice as big in diameter as the rope or cable which will pass through it. If you have access to a power drill put a small hole through the bottom of the bolt portion about one quarter-inch above the bottom. After bolting it to your dredge, insert a cotter pin or a piece of wire

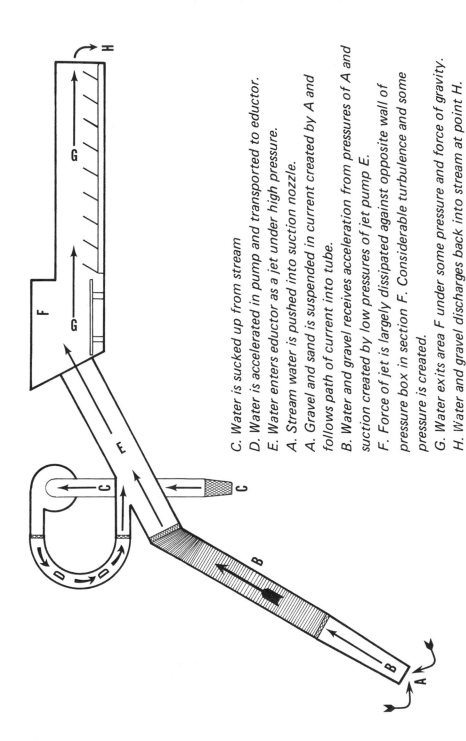

C. Water is sucked up from stream

D. Water is accelerated in pump and transported to eductor.

E. Water enters eductor as a jet under high pressure.

A. Stream water is pushed into suction nozzle.

A. Gravel and sand is suspended in current created by A and follows path of current into tube.

B. Water and gravel receives acceleration from pressures of A and suction created by low pressures of jet pump E.

F. Force of jet is largely dissipated against opposite wall of pressure box in section F. Considerable turbulence and some pressure is created.

G. Water exits area F under some pressure and force of gravity.

H. Water and gravel discharges back into stream at point H.

through the hole and even if your eyebolt comes loose during a long operation the nut will not drop off.

After you have picked a good selection of "handy andy" type of items for your tool box the next consideration is to plan the special tools which will be helpful in the dredging operation. Most important to the dredger are his choices for anchoring lines. Rope is the easiest to use and transport. It is most often used by the fellow who is setting up his dredge for shorter operations. Cable and chain are the best and most permanent and are usually used when the dredge is going to be in one location for several days. Although most chain and cable is waterproof to some degree it is a good idea to rub it with a greasy rag before getting it wet. Never run your hands down cable without gloves and a rag as a snag often develops and can cause a nasty cut.

The final necessary items are the tools of the miner. These might include spades, shovels, picks, rock hammer, hand winch or any other of a hundred items. It all depends on what you expect to find when you reach your dredging area although the shovel, pick and rock hammer are a bare necessity.

Before describing the problems of launching there is one rule that is almost empirical in dredging that must be remembered at all times. There may be cases when it can be broken but only the richest of pockets would cause an experienced dredger to go against this rule. This is simply to dredge upstream and discharge tailings downstream.

Only a little thought about the nature of the discharge is necessary to prove this rule to anyone's satisfaction. The rocks and small boulders of the tailings tend to drop immediately after discharge but the sand and lighter gravels are almost always transported. Since the discharge from the sluice is usually traveling faster than stream velocity the lighter materials enter the stream some distance from the discharge where they quickly settle to the center portion where the velocity is higher than the bottom or surface. This means they can be transported a considerable distance before depositing once more on the stream bed.

If they are discharged above the dredging area, this lighter material can be suspended throughout the dredging area, clouding the water and making it difficult to see where your suction nozzle is working. In most circumstances this also means that the material will have to be re-dredged as barren gravel resulting in lost time. But the most important reason is that the fine sands will almost always be transported to the intake area where they will be sucked up into the centrifugal pump and erode the impellers almost as fast as if you held a file to the spinning impeller. Even a short period of continuous fine sand running through the centrifugal pump can greatly impair its efficiency.

Once you have taken all the previous considerations into account and arrived at the first place you have picked to dredge, make a visual inspection of the general area to determine the size of the area to be dredged. There are two basic types of dredging areas—large and small. The pocket placer is generally considered small. They consist of boulders or other obstructions, areas of eddies or other pressure areas, waterfalls, below rapids, small holes, deep water areas, or almost any fairly condensed area of a stream which can be worked without changing the anchor point.

In a practical situation, selecting the dredging area also means selecting the anchor point. And while there are many variations depending upon the type of dredge, the basic principles are so similar that we will only discuss our now familiar fixed ejector, slightly submerged type of dredge.

If, for example, the area to be dredged is the side of a rock, then you simply anchor your dredge near enough to the rock so that there is plenty of intake hose to reach the depth you want. In this case you would be concerned only about a single, stationary anchor. Once the pocket placer is cleaned out, you will generally move to another spot. These are typical small operations, seldom lasting more than a few hours and do not usually present too many problems.

Big dredging locations are those which encompass more dredging area than can be reached by the practical working length of the suction hose from one fixed anchor point. Both the small and big areas need to have their anchor points determined before dredging can begin, and most types of anchor points can be used for either type. The choice of an anchor and its location are always secondary to the choice of the dredging area. This is because there will always be an area you want to dredge that has no suitable anchor spots. If this is the case, you should not worry for there are methods to accomplish this. Although in picking a dredging area you should always have anchor points in mind, the main concern is the dredging area.

Once the area has been selected there are several other things which must be considered before actually launching and anchoring. One is to make a rough diagram of the area if it is going to be bigger than one or two anchorings. This is best done on graph or quadrangle paper, the kind sold in stationery stores. The diagram should have potential anchor spots, rocks to be cleaned out carefully, or anything else you might consider important. If this is to be an extensive operation of a week or more, you'll find this rough map exceptionally useful.

Not only will it show areas worked and the time required, so an estimate of the remaining time to finish is easy, it becomes a permanent record of your operation. It will show spots producing more gold than others so similar general areas can be worked in preference to low paying areas if time becomes important. If the placer is renewing itself each year, you will have an accurate map to the best potential spots in years to come so that the area can be worked in less time due to eliminating non-paying areas.

Once the dredging area has been inspected and finalized there are three more considerations before actually starting to dredge. These are: point of discharge; point of intake; and anchor points. In a practical operation they are determined at the same time since the selection of any one is also the selection of the other two. But, since each presents its own problems, the three considerations will be discussed separately.

Discharge is usually determined first and the cardinal rule applied: dredge upstream, discharge downstream. In deciding where to discharge the depth of water just below the sluice is important. As mentioned earlier small boulders and larger pieces of gravel contained in the tailings tend to deposit very near the exit discharge of the sluice. Since the current of the stream is seldom fast enough to transport them very far they immediately start creating a small pile which grows in proportion to the amount of material discharged.

If the water is shallow at the point of the discharge this pile of rocks rapidly builds up to reach the bottom of the floats or sluice. When this point is reached two things can happen. One is that a small hydraulic jump is created which elevates the float or sluice and lets the lower end move onto the pile of tailings changing the angle and the sluice velocity. This results in plugging of the riffles and gold values are washed down the sluice and lost. Shortly after the sluice jumps to the top of the tailings pile the pile still grows higher and the discharge and tailings begin to accumulate in the sluice and the pressure box and cause the suction hose to become overloaded stopping the action. This can happen very quickly.

If you are dredging in shallow water and can constantly monitor the sluice this is not too much of a problem as the tailings can be watched easily. On the other hand, if the operator is under water, monitoring the tailings pile is difficult. This is one of the reasons that most scuba operations have two persons with the above surface operator keeping the tailings moving by forcing the rocks downstream using a shovel as a prybar. The problem can not always be eliminated but the time spent in moving the tailings pile can be greatly reduced by selecting a deep spot for the discharge.

Finding a place for the area of intake for the centrifugal pump is always predetermined by the area of discharge since the design of the dredge fixes the position of the intake hose. In this chapter only the basics will be presented and the chapter on operating a dredge outlines more specific procedures when difficulties develop.

If you have been fortunate to find a deep area to discharge the tailings most of your intake problems will be solved naturally. The major considerations are: First, the area should have clearance of at least a foot between the strainer and the highest point of flow sand. Second, there must be a constant supply of water. Generally speaking if the sluice is anchored in three to four feet of water there will be no intake problems and most general dredging areas fall into this classification. Shallow water and bothersome flow sand are discussed in the following chapter.

Once you have determined where you are going to dredge, where you will discharge your tailings and where you will put your intake tube, the point where your dredge is going to be anchored has been fixed. All that remains is to find someplace to anchor the dredge to. But, before getting into attaching the anchor points to your dredge there are two forces besides pressure of the stream flow against the floats that must be taken into consideration.

One of these is the moment of force around the dredge. When any force pushes an object downstream it also tends to rotate it around its center of gravity. Imagine a plank floating down a stream. As long as its progress was unobstructed it would float normally with the forces tending to rotate it over-powered by the velocity of the stream pushing. But attach two lines to the sides of the plank somewhere near the center and it would no longer float down the stream. Now the forces that push the plank would be resisted but those that tend to turn it around its center of gravity would be free to act. If the lines were attached near the back of the plank, the downstream portion might be lifted only a short distance—if they were attached near the center this section could be elevated to a considerable height and the plank would sink at the upstream section. Now the force of the current would push and could, in an ideal situation, exert enough to turn the plank around the axis formed by the lines.

When a dredge is initially launched its center of gravity is well above the water line and near the back section. At this time the turning effect around the center of gravity can be quite pronounced and if care is not taken, the dredge can turn over during launch. There will seldom be any difficulty when the launch is in slow moving water or if the water is no more than waist deep. The big problem occurs when the dredge must be launched in deep water by pulling it across a steadily moving current. In extreme cases, it may be necessary to weight the front end with a large boulder during some launches.

This same effect is very pronounced when the dredge is launched or anchored sideways in the stream presenting its side instead of its back portion to the current. This type of an anchor should be considered as a last resort and a downstream anchor of some sort rigged up to keep the dredge from turning in the current during operation.

Once the dredge is operating, with a full current of water and gravel through it, the additional weight moves the center of gravity downward and the turning effect is not so pronounced as in the launch period. Little advantage is gained by the change of the center of gravity in the sideways anchor and turning is always a hazard.

Not only does the center of gravity change during operation there is another force which helps to stabilize the dredge. This is the force of the jet dissipating itself inside the pressure box which pushes against the forces trying to turn the dredge over. While this is a welcome force in steadying the dredge, it also must be reckoned with in another way.

It is much as if the dredge were floating in the stream and someone constantly pushed it with a jet from a fire hose. If such were the case the dredge would move down the stream much faster than the current because it has so much pressure behind it. When selecting anchor points the combined forces of

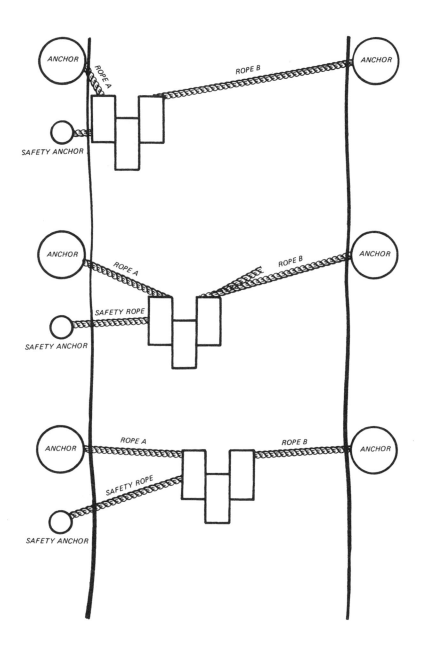

the stream current and the jet current must be considered. Anchor line, points on the dredge and points on the shore to which the line is attached must be strong enough to withstand both forces and not just those of the stream alone.

If both these forces are understood during the selection of anchor points and launching the dredge, there will seldom be any difficulty. Ignore them and you take a chance on having your dredge turn over in even a moderately flowing stream.

Anchor points are usually determined by the natural terrain of the stream and its banks. The best, and easiest solution is to pick two or more rocks which will form a firm wedge on the downstream position holding the dredge rigidly, allowing discharge between them. This is a happy circumstance but nature is seldom so obliging so other ways to anchor the dredge are the rule, not the exception.

The next best solution is to anchor the dredge by means of two ropes on either side of the stream. Ideally, trees, or suitable boulders, will be located on either bank of the stream so that the rope can be safely tied to them. If your vehicle can be brought close enough to the edge of the stream it can serve as one anchor point. In some cases, a large boulder can be rolled down into the general area where it can be used, but this is generally more work than it is worth. Stakes can be driven, or posts sunk into the bank, but this solution is risky because most stream banks are made up of sand which is not a firm foundation for holding a stake or pole. If firm anchor points cannot be found on the sides of the stream, it is usually necessary to resort to stream anchors.

When a shore anchor point is selected be sure that it is firm enough to resist all the pressures anticipated to be working to move the dredge downstream. Next, when attaching the rope (or other type of line) to the anchor point be sure it is firm. The best anchor point is a well rooted tree eight inches or more in diameter.

Tieing the rope to the tree should be done in the following manner. Wrap the rope around the tree once and tie it with a square knot. Then take at least four or five wraps around the tree and then stretch the rope out to the dredge. If you take your additional wraps and tie the knot last the force exerted on the knot will be 100 pounds per 100 pounds of force on the rope. If you tie the knot first and then take five wraps, the pressure on the knot will be about one pound per 100 pounds of force. This means that you will be trusting the holding power of the anchor line to the breaking strength of the rope and not to your ability as a knot tier. It also means that the knot will not be tightened during the full days operation and will be much easier to untie when you are done.

The same principle holds true when the rope must be attached to a rock. Securing a rock as an anchor point is always difficult and depends upon the general shape of the rock. If it tends to let the rope slip up, the anchor line can loosen and fail. Always pick a rock that is well elevated above the stream surface and not one that lets the anchor line run more or less parallel to the surface. Then take your wraps up and let the rope pull down over them.

When attaching your anchor lines to the dredge make certain that they are attached to the frame proper and not supporting bars. Within the limits of good sense attach them as high as possible as this miminizes the turning around the center of gravity simply because there is more weight to lift at the fulcrum which is the point of attachment.

Besides the simple type of anchor which consists of two ropes tied to points on the stream bank, there are two other types of anchors used when the dredging operation is more extensive. Both involve considerable work so while their use is limited there are situations where nothing else will work. Either type can be used when the dredge must be moved frequently either across or up the stream bed. One is a movable anchor, the other is an adjustable anchor.

The movable anchor usually consists of four or five rocks which are secured to rope or chain which is attached to the dredge. It works best with chain which can be bolted to the dredge. It is possible to anchor a dredge with three of these anchors, but it is usually unstable in the stream current. Two movable anchors do not provide enough stability for the dredge as the downstream portion will often move back and forth.

Movable anchors, which are an adaptation of a sea anchor, have so many disadvantages that when you see such a layout being used it is almost always because nothing else would work. Attaching anything to a wet rock is difficult. Some rope, when it is wet and under constant pressure, will stretch and come loose. Nylon rope does not stretch so much but it is so slick under water that it can't be relied on to stay on a rock. Chain can be firmly attached to certain shaped rocks, but on most it cannot be made permanent. Nylon webbing has the same disadvantages as rope of the same material.

Perfect movable anchors can be made by casting cement in a hole dug in the ground with a large eyebolt, or even a portion of the chain inserted in the middle. But when you consider the weight of concrete at about 130 pounds per cubic foot it is easy to see why transporting them any distance is a problem. Movable anchors work well but they are most often manufactured near the site they will be used at and left there when the dredging job is over. They are never used for small jobs.

There major advantage is that these anchors can be used when there is absolutely no other place to attach an anchor line. Such a location might be an extremely wide stream or an area where there are no trees and few boulders. They are fairly easy to move underwater and they can be moved in any direction as opposed to the adjustable anchor line which lets the dredge move only in a straight line.

One of the disadvantages of the movable anchor is that they tend to settle in the sand to a great depth and are often difficult to get out. If they sink in too far, the area around them must be dredged before they can be moved. If you ever have the occasion to use the movable anchor be sure that the downstream anchors are positioned well behind the discharge as they will be buried beneath the tailings which must then be removed before the anchor can be changed.

The adjustable anchor is far more satisfactory for most purposes and it can take many forms. Often it is just the judicious planning of placing two ropes on either side of the stream and shortening one while the other is tightened to move the dredge either upstream or across it. When the area being dredged is straight across the face of a moderately wide stream a line is suspended across the stream and attached to two firm anchor points. Instead of being attached to the dredge permanently the line is run through two eyebolts on the back of the dredge and the dredge can be pushed completely across the stream during the dredging operation.

This method is fast in moving the dredge but some care must be taken to avoid dumping the dredge into the stream by pushing it across the current too fast.

The best material for this type of adjustable anchor is a chain as the links will retard sideways movement of the dredge during operation. Snap on clips of any type can be used to keep it permanently in location in one spot. The next best is aircraft cable and pushing the dredge across the stream is quite easy. Bolt on clamps which are designed to hold a loop in place can be purchased at most large hardware stores and can be attached on either side of the dredge to keep it from wandering away during dredging. Rope works well too but it is a good idea to have two safety lines which can be shortened or lengthened to keep the dredge from moving from side to side.

Movable or adjustable anchors are seldom used except in a two or three man operation which is going to extend much longer than a two week vacation. A lot of thought has to go into their construction and firmly attached safety lines are a must. Remember that the dredge should not be left in mid-stream overnight and provision for getting it to or near the bank is a must.

Once the dredging area has been determined and all the positions of the dredging operation planned, there is one more step that must be taken to insure safety during the launch and peace of mind during the dredging operation. This is securing the dredge with a safety line that is independent of the anchor lines. Although it is not always possible, the safety line should be attached to a dredge component that is not connected to the regular anchor line attaching points. This is so that if these points should fail, the safety line will hold. On the shore it should also be attached to some other anchoring point so if they fail, there is one more chance to keep the dredge from floating away.

Most dredgers use the safety line to bring the dredge to shore at night when shutting down so that the anchor line on one side will still be in position when operations begin in the morning. Often the dredge safety line is smaller than the regular anchor line since it is required only to swing the dredge back to shore in a gentle arc if the regular anchor fails. From a practical standpoint a safety line should never be less than one quarter inch in diameter, and larger ones are better.

Once the safety line has been secured to the dredge the actual launch can begin. There are a lot of ways a dredge can be launched and the method chosen will be determined by the stream current, depth of water, turbulence or other limiting physical factor. The method described here is more elaborate than will generally be used but it is useful for one person in a moderately moving stream that is waist deep, or two persons in a moderately moving stream that is not wadeable.

One writer once suggested that the dredger attach two lines to the dredge, take it to the middle of the stream and then attach the lines to anchors on the side of the stream. This is a good trick if your arms are about 15 feet long. Otherwise it is better to follow the outlined procedure. This is a farily complex launch but if precautions are taken it is not really difficult.

First attach the safety line and allow enough length to allow the dredge to circle back to shore if it becomes loose. Then attach one line to anchor point **A** as shown on page 89. Then estimate how much rope will be required to let the dredge be anchored in the portion of the stream where the dredging is to begin and tie it off at the dredge anchor point.

Next cross the stream and attach the second rope to anchor point **B** on the opposite side. This rope should be at least twice as long as the stream is wide. Bring the loose end of rope **B** back to the dredge and run it through the other eyelet.

Get a firm grip on the dredge with one hand and grasp rope **B** with your other hand pulling it just taut. Now you can start wading and guiding the dredge gradually to the center of the stream. As you move out, keep taking up the slack of rope **B** so that it maintains a slight tension on the dredge and helps fight the current.

If rope **A** will become fouled on rocks or vegetation during the launch, it can be coiled on the dredge and allowed to play out gradually.

Once the dredge is in the position you have selected for dredging, rope **A** should be fairly tight and rope **B** can be tied off. The surplus line from rope **B** can be coiled and tied around the line giving additional safety. Once this has been done the safety line can be tightened and any adjustments to the anchor points made. Now the dredge is ready for starting the engine.

Naturally the above procedure can be much easier if two persons are present. However, one person can safely manuever a dredge to the middle of a moderately moving stream if necessary.

If the dredging operation is going to be done by scuba diving the stream will probably be too deep to wade and some modification of the procedure is necessary. The operation in this case can be done by two persons, but three are better. The main problem is that it is practically impossible to start the engine in water that cannot be waded. Therefore the engine must be started and a jet established before the dredge can be actually launched. Starting the engine usually takes place near the shore where the water can be waded. The jet is established and checked to see if everything is working right.

With the dredge operating, moving it is more of a problem and a third person to steady the downstream side by swimming along is very helpful. In any case, two persons are necessary. One handles rope A and plays it out gradually keeping it taut until the dredge is in the dredging position. The second person stands on the opposite shore and pulls on rope B until the dredge reaches midstream. This can be done by having rope B pass through an eyelet and returning to the bank, or, as is more often done, tieing robe B to the dredge and later tieing it to the anchor point once the dredge is in position. When the anchor lines are tight and the safety line secure, the dredging can begin.

Regardless of the type of anchor being used, it is not a good idea to leave a dredge anchored in midstream overnight. Beaching the dredge is just the reverse of launching it and depending upon the width of the stream and the velocity it can be done by one person, but two are better.

If the stream is wadeable, stop the engine and beach the dredge by walking it to shore. If this is not possible, beach the dredge with the engine running (in most deep water operations the dredge must be beached each time gasoline has to be added). To do this rope B is usually untied at its anchor point and allowed to play out slowly letting the dredge pull against rope A until it beaches on the opposite shore after following a gentle arc through the water. If another person is available, he can keep the safety rope tight at all times which will tend to steady the dredge on its way to being beached.

If the dredging operation is going to take place in the same area the next day, the launching can be made quite simple by pulling on rope B until the dredge is once more in the center of the stream.

Most launching operations are much simpler than this and few are harder. The secret to successfully positioning your dredge is to take nothing for granted and to plan every step in advance. It is also well to remember that even a moderately fast mountain stream can sweep you off your feet, especially when you are trying to maintain the balance of your dredge as well as your own. It is best to be wearing a life jacket or similar safety device during a complex launch to be on the safe side if something like this should happen. Actual operation of a dredge does not require a safety jacket unless the current is quite swift.

12

DREDGING PRINCIPLES

By now you should be familiar with the fundamental principles of your dredge, the methods of getting the water running; and the best ways to launch. All that remains is to start dredging for gold. It might seem strange that so much material would be covered before the actual process of dredging, or more specifically, operating the suction nozzle, is covered. But I feel that anyone can only concentrate on dredging if his rig is working properly. Now that it is, and the dredge is safely at an anchor point, let's get on with the dredging.

While telling you many hints for dredging, I'll occasionally throw in a bit of theory so that you will be able to solve the problems I do not have the space to cover in this chapter.

For your first dredging operation try to find a patch of stream bed that is well classified and mostly sand or light gravel. Earlier I described the current that was created around the nozzle by the pressure forcing water into the tube. It is an often stated rule of hydraulics that pressures in a fluid work in all directions equally. Therefore, the currents around your suction nozzle will be pushed toward the intake as a circle several inches bigger in diameter than that of the suction nozzle.

At first hold the nozzle an inch to several inches from the sand being sucked in and observe how much sand and gravel is being moved. As a further experiment, put the nozzle very close to the sand and hold it there. Notice how the gravel is sucked out in a hemisphere shape. Held this close to the sand the nozzle pulls in less water and less gravel. If you keep pushing the nozzle in this small depression, the hole will become smaller and eventually close around the suction nozzle causing no sand or water to be sucked up because the gravel forms an effective dam against the water.

After trying these two tests you should be getting a good idea of how successful dredging takes place. There is a spot anywhere from a half-inch to a couple of inches back from the sand that your dredge will suck in the proper amount of gravel in proportion to the water being moved up the tube. Determining the proper proportions of water and sand is difficult and is usually a judgment decision based on the type of gravel, the degree of cementing and other factors. The percentage of gravel to water is very small, most engineers estimate it at five to ten percent. The limiting factor as to how much gravel can be picked up is the amount the sluice can handle without plugging.

Dredge manufacturers often rate their various designs at so many cubic yards per hour. These figures are obtained in a controlled test with well classified gravels which usually contain no boulders. The gravels you dredge will seldom fall into this range and the efficiency of a dredge in the field is usually somewhat less than the rated capacity. Another misconception many dredge owners have is that their dredge is not going to work as well at 10,000 feet altitude as it does at sea level. Actually the dredge will work just as much gravel at any altitude provided the engine fuel/air mixture is adjusted to compensate for the reduced air pressure. There some effects of air pressure, temperature and gravity that reduce the efficiency but these are so small they can be ignored. A car that goes 60 mph at sea level will also go 60 mph at 10,000 feet if the carburetor is adjusted. The same holds true for dredges.

A few minutes of operation will give you the best distance your nozzle should be from the gravel. It is a good idea to also check the flow of gravel through the sluice to make certain it is not plugging or washing out of the riffles. If either of these are occurring you can compensate by changing the speed of the engine, pulling in less gravel, or, on some dredges, raise or lower the angle of the sluice. Once you are satisfied that the gravel is moving through the sluice properly the dredging operation has begun. It is as simple as that. All the preliminary work and homework actually consumes so much time that most persons don't realize how simple dredging really is. It takes some work to actually dredge but most of the effort is really consumed bracing your body against the current. Moving the suction nozzle around requires little effort as it weighs very little under water.

Once the gravel is flowing through the sluice there are only a few general rules that need to be observed. One is to forget all the rules that anyone has told you about following a pattern of suction on the stream bed. As soon as you begin cleaning out your first pocket placer you will soon discover that the pattern the suction nozzle makes is going to be determined by the type of pocket or area you are placering. If you dredge straight across the stream bed, for example in three anchorings, you will probably be moving in a semi circle if the stream current is filling up your cavities with speed. If the water is flowing relatively slow and the dredged hole is not filling up fast it may be possible to remove overburden all the way across, then switch to a different type of suction nozzle and clean up the better paying gravel near bedrock.

There are two basic methods for cleaning out placer deposits. One is the quick descent used almost always in pocket placers and often in larger ones when flow sand fills the hole quickly. In this type of dredging you excavate a circular area about three times bigger than the area you want to dredge at the bottom. This is a very hard estimate to make because the proportion of the surface excavation to the bedrock excavation depends upon how fast the hole is filling in and how deep the excavation will be. The general idea is to dredge to bedrock, or the bottom of a pocket placer, before the hole can refill with gravel.

The second method is to excavate large areas of overburden, usually with a greater capacity suction nozzle, and then return with a more efficient suction nozzle as the depth nears bedrock. This method is most often used by divers and, since it is such a lengthy process, only in bigger operations. No preset rules can be given as the size and nature of the excavation depends on the overburden, current of the stream, nature of the gravel and many other limiting physical factors.

Defining the nature of an excavation is difficult but there are a few other procedures that are easier to explain. One of these applies to boulders. Most writers, and many dredge manufacturers recommend removing boulders from a dredging area. I take a different view point and believe that winching a boulder out of a stream is a last resort. This great passion for moving boulders from a stream seems to be a holdover from the early days of hydraulicking, dredging and placering. In hydraulicking the loosening of a two ton boulder high on the side of a hill being hydraulicked was an awesome thing. With the slope lubricated with water, the boulder rapidly gained momentum and if it traveled far enough to reach the sluice its mass could wreak havoc destroying many dollars worth of equipment, or pose a hazard to the miners. As a result when a large boulder was located it was usually blasted, whether on the ground level or high on the side of a hill. The same held true for bucket dredges. These, too, could lift a two ton boulder and then drop it on a fragile sluice or jigging system and prior removal or blasting was the only way. The small time sluicer, however, looked on removing boulders as just a lot of unnecessary work or expense and worked around them. Only if it was absolutely necessary did he remove them and this is the way the modern gold dredger should look at boulders.

The main reason for ignoring boulders if possible is simply weight. Let us consider three boulders; one will be a sphere 48 inches in diameter; the second a rectangle 12 by 24 by 48 inches; and the third a relatively flat one, six inches thick by 30 inches by 60 inches. There are few rocks that are this perfect, but on the other hand there are many which come close and besides the weights on these can easily be computed so a person can get an idea of what he might be facing in removing them. Let's use a weight of 170 pounds per cubic foot, about the density of granite or gneiss.

A quick calculation reveals that the flat stone is the lightest at 1,062 pounds; the rectangle weighs 1,360 pounds, but the sphere will always fool you at 5,695 pounds. This is a good example of how judging the weight of a boulder can be tricky. But there is even more to be thought of. We must consider how much these boulders weigh under water because this is the place we must move them from. Respectively the weights would be approximately 670, 860 and 3,600 pounds. It is easy to see that if the stone was being lifted out of the water by a winch or crane the sudden addition of the weight as it broke water would be considerable—a little over a ton in the case of our imaginary spherical boulder. Even when the force needed to move these stones is reduced by inclining the angle of pull, the work and time consumed in moving just one can be considerable.

If boulders came by the ones, it might not be so bad—but in streams they come by the dozens and the fellow who goes dredging with the idea of moving them all is going to do nothing but move boulders. It is far easier to work around them—or even better—to work them down or to the side so that the gravel beneath them can be exposed and dredged. It's faster and with a little caution, quite safe.

Under water a boulder with a size of one cubic foot will weigh around 107 pounds. It can often be moved with a prybar but the depth often makes even this difficult. Usually stones of this size, and larger, can be moved by dredging on one side until a depression or hole has been created. Then simply slide or push the boulder towards the hole and it will naturally roll in. This procedure can be repeated on every pass and eventually the boulder is at bedrock, which is where it was headed in the first place.

Large boulders are more of a problem. To weigh a ton a boulder only has to be a cube with sides of about 27 inches in length. When boulders get huge they can weigh many tons and are often beyond the capacity of even larger, motorized winches to move. These big boulders are a nuisance and sometimes a hazard. Dredging beneath one can sometimes lead to disaster if care is not exercised. Fortunately, the stream tends to deposit a big boulder in a position where it is most unlikely to roll down the stream—this is what causes it to come to rest during the process of saltation. But this rule cannot always be relied on. If there is any doubt as to the shape of a boulder under the stream bed, clean out beneath it with caution and leave enough gravel on the sides to keep it from rolling.

After boulders the second most bothersome thing is maintaining the quantity and quality of water for the intake hose. Since the suction nozzle keeps getting deeper in the water, quantity is seldom a problem here.

Estimating the quantity of water required by the centrifugal pump alone is difficult because most persons cannot relate gallons per minute to a really definable shape. Imagine your 150 gpm pump hooked to the 20 gallon gas tank of your car. It would pump it dry 7½ times per minute. Or, if you had a barrel 2½ feet in diameter and four feet deep you would have to fill it once each minute to keep your pump operating. That's a lot of water and it does not stretch anyone's imagination to realize that a small hole in a slowly moving stream is not going to provide enough water for some dredging operations.

In the field imagine a stream of water flowing by the foot valve as 1½ foot wide and two feet deep. A float in this stream should move about seven feet in one minute just to supply a pump with a capacity of 150 gpm.

Quality refers to the amount of sand suspended in the water being pulled into the centrifugal pump. Ideally there will be none but practically speaking there is usually a small quantity in any moving stream. Naturally, any sand passing through the centrifugal pump will abrade its impellers. Most pumps can tolerate minute amounts of sand and work efficiently for years. On the other hand, large amounts can wear down the impellers quickly—sometimes in a matter of a few hours. Avoiding such damage is largely a matter of maintaining the quality of the intake water.

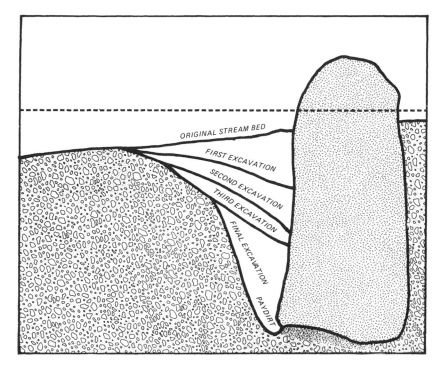

Dredging down an obstacle such as a boulder is ususlly done in a series of passes. The widest excavation is done first, the final is often just large enough to work in.

Diversion dams are tricky. Divert no more water than absolutely necessary. Dam on right is made of fence to keep out vegetable matter. Middle one is to slow down stream flow to keep sand out of the intake nozzle. Obstacle on right is a natural dam.

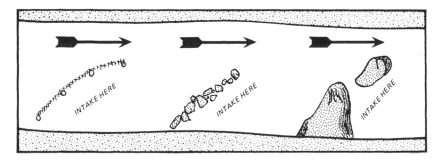

First, all tests for quality must be made with the dredge in full operation. Occasionally natural movements of the intake nozzle causes sand to be suspended and transported to the nozzle and this cannot be noted unless the suction tube is pulling up sand in large quantities. Simply observing the water around the strainer is not a good enough test. Even in the clearest of mountain streams, light colored sands are often impossible to see. In some cases it is possible to put your hand in the water near the strainer and feel the abrasive sand by rubbing your thumb over your fingers. If a more positive test is required use a clear drinking glass to obtain a sample of the water. Hold your hand over the top of the glass and put it near the strainer. Remove your hand allowing water to run in and then close the opening again. Bring the sample of the water to the surface and let it set for a few minutes and the sand will settle to the bottom. If there is a large quantity of sand, the impellers stand a good chance of being damaged.

Flow sand is present in virtually all streams and is generally heaviest in a space of a few inches near the bed. In some streams the flow sand can work its way up into the faster moving center portion of the stream. This is why most dredge manufacturers make their dredge intake hoses so that they will reach only a few inches into the water. Usually the strainer position is well above the danger level for flow sand but yours may be just long enough to reach a portion of the stream where sand is flowing.

If this is the case the problem can often be eliminated by a float-anchor arrangement. A rope is tied to a float or somewhere on the dredge frame and also to the strainer pulling it up out of the flow sand area. In this position the strainer may be so high that it might pop out of the water occasionally causing a lack of suction. To avoid this another short piece of rope is tied to the foot valve with the opposite end tied to a boulder which serves as an anchor to steady the strainer in one position.

Some dredgers get very elaborate when flow sand or vegetation becomes a problem. Vegetable matter does not always abrade the impellers but it can plug up the strainer or intake hose reducing the pump efficiency. In most cases they build a dam. This might work in some special cases but more often the cure is worse than the disease. In the first place, diversion dams are hard to build on short notice. Second, if it is an underwater dam(with water flowing over the top) it will probably create a small hydraulic jump on the upstream side and eddies at the bed on the downstream side. This will agitate the bed at the base of the dam and create more flow sand than there was in the first place. A diversion dam should be considered only as a last resort. If built at all it should angle downstream so that the "riffle" effect of the dam will be minimized.

Since the only real practical use of a diversion dam is to eliminate large quantities of vegetable matter such as leaves or plants there are many other, easier ways to solve the problem. One way is to construct a second strainer out of quarter-inch screen and install it over the regular strainer. This will not inter-

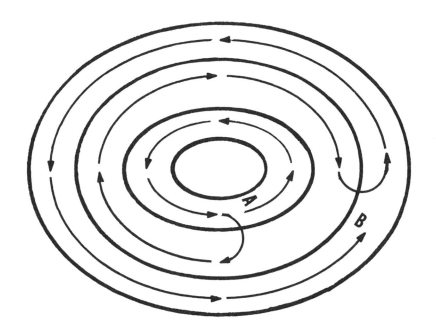

Circular areas are usually cleaned out in a circular method, starting
with the innermost area. The common pattern is to follow the
arrows starting with A and proceeding to B.

Larger areas are usually cleaned out in a systematic method of
enlarging the holes, whether the excavation is rectangular or
circular. The first excavation is the shaded area. Then the sand
is removed from Section A. Next the shaded area, then Section
B and so on until cleaning out a shaded area reveals you are
at bedrock. Then extreme care is taken in the sluicing for
this is where all the gold is.

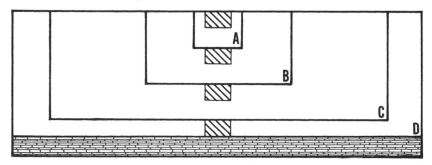

fere with the suction and can be cleaned of leaves without any trouble. Or, if large quantities of vegetable matter are present, it is possible to divert it by constructing a "dam" using the same type of screen. If angled downstream it is almost self cleaning and does not create turbulence as does a normal dam.

Another method used to maintain the quantity and quality of intake water is to put the intake hose inside a bucket or barrel which is anchored or set into the stream bed. This particular operation is also used in shallow water operations but it should be used with extreme caution, for like a diversion dam it can create more problems than it solves.

If the bucket is small—say five gallons— it will be emptied and filled about 20 to 25 times per minute. The rapid flow of water into the bucket will create unseen turbulence which can cause water vapors which will be pulled into the intake hose. This causes a drop in efficiency and can cause cavitation which is almost as damaging to the impellers as is sand.

There is another problem with the bucket or barrel used to increase quantity or quality of intake water. This is that it might become a natural trap for the flow sand it was intended to eliminate. The barrel becomes a trap when the top opening is placed near the level where the flow sand is being transported. Water entering the barrel does not "fall in from above" as might be first suspected, Rather, a low pressure area is created inside the barrel because the flow into the barrel is at a higher velocity than that of the stream. Since water under pressure moves in all directions with equal pressure, not only the water above the barrel flows into it but water along the sides is pushed up to join the flow to the barrel. If the opening of the barrel is near the bed, flow sand can be entrained in this upward flow where it enters the barrel and is swept through the intake tube to the centrifugal pump. If the opening of the barrel is very near the bed, a quantity of sand much more than that of the normal flow can be entrained and damage the pump more than if the barrel had not been used in the first place.

Besides maintaining the quantity and quality of the water the dredger must maintain the flow of sand through the tube. There are two ways the suction tube can become fouled enough to stop the dredging process. One is by overloading and the other by obstruction. Since overloading is the easiest to solve, let's look at it first.

Overloading occurs when the weight of the material being dredged becomes too great to permit lift and efficient sluicing. In waist high depths it seldom occurs but in diving to depths beyond six feet it starts to happen more frequently.

The first indications of overloading are usually detected at the suction nozzle and not in the sluice. Suction at the nozzle is greatly reduced and the percentage of gravel becomes more than the dredge can effectively handle. Carried to its ultimate, overloading can cause cavitation at the base of the jet. There is probably no quantity of sand that can be so great that it could change the pressure advantage at the nozzle so that sand flows backwards through the lower portion of the suction tube. What happens in overloading is that sand and gravel creates high frictional forces inside the tube which can become so great that the movement of the stream is slowed. Sometimes this effect can be quite pronounced and the suction at the nozzle almost quits. The problem is found more in the fixed ejector near or above the surface type, and almost always occurs at depths greater than ten to 15 feet.

The solution is quite simple—merely back the suction nozzle away from the gravels and allow more water to flow up the suction tube until the ratio of water to gravel is increased. The velocity inside the suction tube will rapidly accelerate and gravels will start moving again. You will be able to tell by checking the suction at the nozzle. It is an indication that you have been trying to suck too much gravel up the tube and to avoid it happening again just increase the distance of the tip from the gravel by a small margin. Once you can see the gravels moving into the tube regularly at the same approximate speed, you know that you have found a good suction distance for the type of material you are dredging.

There is another type of overloading that is more difficult to detect by the suction tube operator because it does not always show itself in greatly reduced suction. It happens when the overloading occurs between the base of the jet and the pressure box. Since the base of the jet will always create considerable suction, sands and gravels will continue to flow in but they will not be accelerated as much as the jet travels back up to the pressure box. With greatly increased sand in proportion to the water, plugging of the riffles will occur very quickly and all the remaining values will be washed through the sluice.

The answer here is basically the same. Flush the suction tube with water and when the gravel starts moving fast enough, start dredging again but not in such great quantity. This type of overloading occurs most frequently when the ejector is located below the surface and seldom does overloading occur in the first ten feet of depth.

Overloading is particularly troublesome with the underwater sluice. Due to the nature of its operation, the diver seldom checks the riffles until it is time to dump them. If the suction tube has been overloaded he may waste an entire 30 minute shift doing nothing but running sand over plugged up riffles. It is a wise idea to check the riffles of an underwater sluice shortly after beginning the dredging to see if you are overloading your suction tube for the type of material you are dredging. It is only after considerable experience that the operator of an underwater outfit can become proficient in estimating the amount of material to be sucked up at any given time. Due to this problem, most underwater sluices are used today only to clear off excessive overburden where low values may be expected. Close to bedrock the dredger often switches to the surface sluice where these problems are not so great.

Since overloading occurs at greater depths, the average dredger will seldom experience them. His problems will more often be those of too much velocity and water in the sluice causing materials to be washed out.

One problem plagues everyone and this is the out and out obstruction. Obstructions are sometimes not so easy to eliminate but generally speaking they cause relatively few slowdowns. Almost any solid matter can cause an obstruction—matted vegetable matter, sticks, boards, etc., but most obstructions are caused by elongated stones which enter the suction nozzle by presenting their narrow side and then twist in the tube to become wedged.

The first thing to do is to try to isolate the general location of the obstruction. If it is above the jet, the water supply to the sluice will be greatly reduced, suction will slow down or even stop and in some cases, the obstruction may be so large it will shut the flow of the jet down, reversing it and the water will flow out of the suction nozzle with tremendous force.

If the obstruction is below the base of the jet, water will continue to flow to the sluice although its quantity will be reduced. The water flowing through the sluice will contain little or no gravel. Depending on how much of the suction tube is blocked, suction at the nozzle will be greatly reduced or cease to exist.

The solution for either type of obstruction is to clear the suction tube by removing the obstruction. Usually this can be done by getting it in the same position it was when it entered the tube and letting it be transported to the pressure box.

This is relatively easy if the obstruction is in the suction nozzle or the flexible portion of the tube itself. Lift the suction nozzle underwater until it is pointing upwards. If the obstruction is here it can usually be seen and moved with a short rod. If it is in the flexible part of the tube it can be changed in position by lifting the tube in the general area and gently twisting it. If this doesn't work try tapping the tube several times along its length to see if a little vibration will move the obstruction.

If the obstruction is near the suction tube orifice past the base of the jet, it must be removed from the other end. Most dredges have a plug in the pressure box. Remove the plug and insert a long rod (most dredges come with this as an attachment) poking down the throat of the ejector. A surge of water and gravel will tell you when the obstruction is moved.

In removing either type of obstacle it is best to have only water being sucked in through the suction tube. If the nozzle is left where it can suck in gravel, smaller stones will pile up behind the obstruction and quickly create a major plugging of the tube that requires a lot of work to clean out.

If neither of the preceding procedures work, the next best thing to do is shut down the operation and disconnect the suction tube from the ejector and manually remove the obstruction. Don't use a rod longer than the metal ejector tube. Invariably it will slip to one side and puncture the suction tube. There is really no satisfactory way to repair a suction tube and many feet of expensive tubing can be ruined by one slip.

13

PRINCIPLES OF LODE PROSPECTING

Looking for lodes follows many of the general rules of placer prospecting but, since the search is made on dry terrain, there are several other considerations.

The search for a lode can be divided into three different segments: 1. Deciding on a localized area or starting point; 2. The tracking search; and, 3. The lode or source search.

The first step is deciding on the localized area. Once this has been done the prospector begins looking for signs or indications that a precious metal, usually gold, exists.

Deciding on the localized area is done in many ways. Sometimes the area selected will be a stream that has been good for placering. Occasionally a gold dredging operation will turn up other valuable minerals such as the platinum group or precious stones and the prospector will want to find their original source. In this case, the stream becomes the starting point.

On other occasions, and this is the usual case, the trip will be made to a localized area on the assumption that there should be a valuable mineral in the general area. The assumption may come from a study of old documents, source books, geological maps or the past mining history of the area.

In this second situation the prospector will have no starting place and must search for the first indication that gold or other valuable minerals may be present somewhere. The most important consideration in this type of initial search is to look at the lower elevations of the drainage system where the valuable mineral is expected to exist in lode. Gravity, water and prevailing winds should all be taken into account in this initial investigation.

The third way a localized area is found is the kind that generates the most excitement and usually the most frustrations. This is when the other two searches are eliminated and the prospector has the lucky find of a rich piece of float. Sometimes it is his own discovery but more often a friend or casual acquaintance makes the find and reveals it to the prospector through friendship or casual conversation. The most important thing is to discover the exact spot where the float was discovered. Generalizations are of little use and the description "near the old Boulder mine" can mean so many things that the location of the lode will never be found. Try to get a precise location or area. The best method is to make a personal visit with the finder. Next best is to have him point out the location on the smallest scale topographical map available. This spot then becomes the starting point.

Before any search can begin it is necessary to define the word "float" as it applies to prospecting. Float is any isolated piece of mineral or ore that has been broken off a body of a vein or outcrop and which has been transported by natural geological agents some distance from its original source. One of the major difficulties in locating the original source of float is that it can also be transported a considerable distance by forces other than geological agents.

This is well illustrated by some of the very rich pieces of gold that were found in California's desert. In the early history of the area, native Indians often picked up rich pieces of gold from weathered outcrops simply because they were pretty and perhaps they intended to make some form of jewelry out of them. Later, perhaps on a long trip, they would be discarded because the rocks became too heavy to carry. Later these pieces of float were discovered and caused prospectors to spend countless hours looking for the source in the wrong drainage pattern.

Other methods by which float often got transported to another drainage pattern were from being dropped off of ore wagons while the ore was being hauled to a mill; from saddle bags; lost by another prospector who was either killed or who simply failed to return to a cache.

One geological agent which can transport float miles and also across drainage patterns is the glacier. When settlers first arrived in portions of the middlewest the Indians had ornaments and tools made of various soft metals which were almost pure so they could be worked by hand. Concentrated searches for the source of these minerals revealed nothing. At first historians believed the metals had been obtained through trading despite local legends that said they were found. Later geological studies showed that this area had been the terminus of pre-historic glaciers and most geologists now attribute the occurrence of these pieces of float and Indian artifacts to this source.

When a piece of rich float is found it is well to compare its identification with the other ores and minerals which have been found in nearby areas. While finding similar circumstances many miles away will not rule out its occurrence in the local drainage pattern, this fact can be of use in deciding to abandon a search after much time has been put in on a fruitless search. Also compare the makeup of the float with the local geology. This advice should be absorbed with caution. Discovering gold in white quartz float in a country where there are few white quartz veins does not preclude there being a white quartz vein hidden someplace in the search area.

One important fact should be emphasized. Float, in itself, may not assay high enough to be of commercial value. This should not rule out a search on its own. The float may lead to a lode which is richer, or because of its size and ease of mining, is commercially exploitable.

There are three common ways for float to be discovered. Most often it is found laying on the surface where it was recently deposited after a short movement. Surface float is also found after it has recently been uncovered by wind or water erosion.

Secondly, float is found in a stream, gulley or wash where it has found its way and been transported further by water. This type of float may be either a rock or a well defined placer deposit consisting of the mineral itself.

The last place float is commonly found is in a placer deposit itself. This differs from situation two in that the placer consists of another mineral and traces of a second mineral are located, usually at the time of final concentration. A good example would be minerals of the platinum family which are often located with gold placers. Generally speaking, it is true that if one mineral occurs with another in placer, they will also occur together in the lode. But occasionally both will have separate sources and this adds to the problem of finding one or both lodes.

Float can occur in many sizes, from minute grains which must be discovered under a strong magnifying glass to huge boulders. This is the reason for one of the more or less true general rules of early day prospectors. The bigger a piece of float, the less distance it has traveled. If the drainage system is small, perhaps only a few square miles in area, this rule can generally be relied upon. But in extensive areas its validity is questionable. The geologist tells time in millions of years and even a large boulder can be moved many miles in this period.

Another general rule that can more often be relied on is that the sharper the angles of the float the nearer it is to its source. This is frequently quoted to indicate how far placer gold is from its original source. Well rounded grains indicate long travel, the sharper ones, less. Once again there are so many variables that the rule can be false as often as it is true. Slow moving streams may grind and smooth gold particles for a considerable time before it is moved any considerable distance leaving them rounded and still close to their source. On the other hand, pieces of ore or grains of gold can be moved a long distance from their original source during times of flood where they are deposited deep, never to be smoothed again until they are recovered by a prospector.

These two rules are good ones and size and angularity should be given some consideration in planning a lode search. However, the more successful prospector will be the one who takes the exceptions into account when he makes his tracking search.

If the float has been located in a moving stream, wash or gulley; or if it was located in a stream placer, this becomes the first starting point to find the true starting point. In this case the initial search starts upstream following the procedures outlined in the chapter on placer prospecting. The exception to normal placer prospecting is that this is a search for "sign" and not pocket placers. The lode prospector seldom stops to clean out a placer unless it is just too rich to pass up.

If the stream has moving water the search is conducted with a pan or very small dredge. This lightens the load the prospector must carry and reduces the time it takes to set up to sample a likely looking spot.

While he doesn't stop to clean out the better placers he finds, the smart prospector will carefully mark the location of each promising location and return here with a larger dredge when time permits.

Strangely enough, what the prospector is looking for is the point where there is no gold at all. There is no way to prepare for this result and constant sampling must be done. Depending upon the terrain, the overall length of the stream, water conditions and other factors, the distance between sampling areas can vary from a few hundred yards to miles. The most important thing to remember is to use basic stream placer prospecting techniques and look in the most likely place where heavier minerals will be deposited.

When one or two areas are discovered that show no gold at all when sampled at depth, he retraces his steps going downstream. The prospector backtracks to the last point he found colors. This location should be somewhere between the no-color location and the spot where the last colors were recovered.

If the colors peter out immediately above a tributary stream, this stream is then prospected upstream to see if it shows color. If it does the same procedure is followed until the color runs out.

At some time in the stream search procedure the color will disappear at a point where no stream connects. An immediate search takes place above the area on both the right and left banks of the stream to determine where the gold has been originating. This search may have to go on for some time upstream since there is no guarantee that the gold has been transported, but not deposited for some way.

Both banks should be examined closely for it may be that the stream actually is eroding a lode or permanent placer somewhere in this area. Tertiary river beds, bench placers and even veins are cut by streams and this may be the original source.

Once this possibility has been eliminated the area where the gold is determined to enter the stream becomes the true starting point for the tracking search.

The overall search purpose is to find a lode but in prospecting for gold it often leads to the discovery of preserved, permanent placers such as a Tertiary river bed. Technically this is not a lode but since the process for discovering such a deposit is exactly the same as that for finding a lode, no distinction will be made between the two here.

The first step of the tracking search is a study of the ground slope to establish a drainage pattern. This is best accomplished through a topographical map. If the map is also a geological map of surface and sub-surface features such as fault locations the type of formations, general mineral sites, so much the better. The geological map can establish not only the drainage pattern but contacts between different formations and faults and these are one of the most suspect areas to search first.

If no such helps are available the search begins up hill in the drainage pattern. There are no precise rules for every search, but a systematic triangular type of investigation can be applied to almost any type of prospecting endeavor,even many of those conducted through geophysical methods.

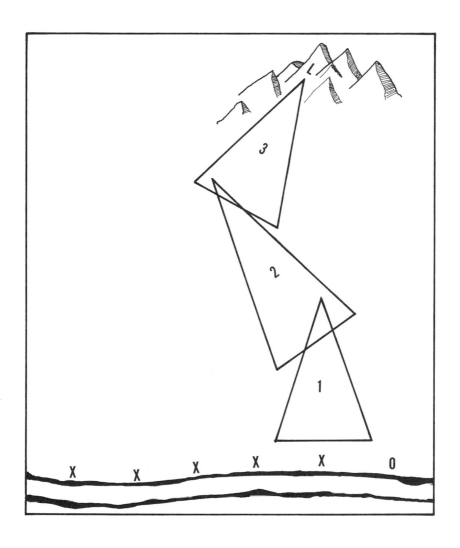

Generally speaking, a lode search follows a stream (running or dry) until the gold samples peter out. Then the search is made on the surrounding terrain following the general pattern of the watershed towards higher ground. In this instance, the X's indicate that a sample of the river showed gold. The O shows the spot it disappeared. The triangles varied three times and the L indicates where the lode was found. Lode searches are seldom this complex and most involve only one triangle.

To begin the tracking search locate the highest point from which the float could have originated. This can be done either in the field or with a topographical map. If no map is available it is a good idea to draw a rough map at this point. Next draw a line from the place where the float was found (or the place it is determined that the gold petered out in a stream) to this high point. This line is called the *search line.*

Theoretically, the float should have followed this line of travel but it is doubtful if any float was ever so obliging. Therefore it is necessary to construct a search area which will include the most likely spots the float could have originated but at the same time to keep the search to a manageable level so that time will not be wasted.

To establish the perimeters of the search the prospector stands at the bottom of the search line and draws a line perpendicular to the search line in both directions. The end of this line in either direction is the point furtherest away that he estimates more float could have been found. He marks these on his map and connects these lines with the high point.

A glance at the illustration accompanying this section will show exactly how a search is conducted. It would be a mistake to think of this search area as a perfect mathematical triangle. On some maps the legs of the search triangle are going to be as wavy as a piece of cooked spaghetti. The general idea is to isolate a search area and investigate it systematically. It will occur to most experienced prospectors that this is what they have been doing instinctively all their life. It is the natural way to search for anything that lies hidden on the side of a slope or hill. Since it is a very natural thing to do, few persons ever make a map or mark a topographical map with the triangle. This is a mistake for if the first search is unsuccessful, this area can be eliminated from future searches; or it may show a new starting point; or it can reveal much about when to start in a new direction and what that direction is.

From the on the site inspection, or from the topographical map the grade can easily be classified as steep or gentle; the search line as short or long. These factors will change search distances and times spent on the search but they will not change the basic principle of establishing a base line and systematically searching each base line. Also the terrain may indicate that at certain points new search triangles must be set up to explore gullies, canyons of a mountain or small rounded hills. In the case of the latter the prospector may conduct a circular type of search using the contour base line principle.

Now that the prospector has drawn his search triangle either mentally or on a map, he returns to the point where the base line is intersected by the search line. Sampling is done along the base in regularly spaced locations. In the case of a base line of several miles and a gentle slope the space between exploratory sample sites might be several hundred yards apart. If the search is in steep terrain and the base line less than a mile the sites might be only a few yards apart.

As these samples are taken the prospector should note on his map the results of what the sampling determined. This can be in any language or coding he chooses. He could grade the samples on a scale of from one to four, or simply put down zero for no colors and "C" for some colors. The weight or amount of colors could be recorded but this might tend to get confusing as a rich piece of float could throw off the averages indicating an erroneous pattern. Personal experience and preference will determine how the test holes are rated and the only thing necessary is to grade each prospect hole with the same criteria. Once the prospect sites on the first base line have all been examined the prospector will move to his second base line.

While the first base line of the search is almost always fairly straight the succeeding lines will follow the contour lines of the terrain and will soon lose any resemblence to a straight line.

The second base line, and all succeeding ones, is determined from a higher contour line. This is easily drawn from the contour line on a topographical map. (Note that it is not always the next higher line that the mapmaker chose. The prospector uses his own judgment of the next elevation and may skip three, four or more map selected contour intervals.) If no map is available for the area, a hand level or similar instrument can be used to determine the next elevation to be explored. Contour levels should be as accurate as possible but do not need to be laid out with the precision a surveyor would do.

As mentioned previously, finding the distance between base lines is determined by the slopes of the land and the prospector's judgment and experience. In gentle slopes the terrain may rise only a few feet per mile and the distance between base lines as much as a mile or more. With a short search line and steep terrain, the slope may rise ten feet in every 40 yards and this would be the next base line. The infinite base line would be the bottom of a cliff where the search would be made at the base and the next search at the top.

Once the second base line has been determined the same process of sampling is done as was done on the original base line. The spacing of the holes will become less and ideally should creep inward in the same proportion as do the outside legs of the search triangle. Don't worry about this too much, though, as it has little to do with the end result.

As the search progresses and the base lines get shorter it is possible to systematically search more lines in a shorter time. But the two cardinal rules of the search triangle should never be broken. First: Always search on a contour line and then move to a higher contour line for the next search. Don't jump back and forth between elevations before a base line search has been finished. Second: Do not skip base lines indiscriminately. Unless there is a reason for skipping the base line due to topography, impassable objects or a new triangle based on fact, each predetermined base line should be explored.

At times it may be necessary to change the search line and start a new search triangle based on fact. This is where the importance of keeping track of how the sample holes produced becomes evident.

It sometimes occurs in a search that the wrong high point has been selected for the search. Often the original source of gold is at a lower high point. This is especially true if the original search triangle was very large covering hundreds of square miles as might be experienced in some places in Alaska.

As the results of the sampling tests are recorded systematically on the map, a pattern often begins to show up. This pattern does not always follow the search line and often cuts it at a well defined angle. When such a pattern is developed, it is time to establish a new search line.

Interpreting these patterns can be an art in itself and if the prospector has any doubt as to the pattern or angle of accumulation he should take another group of samples from the next regular base line to firmly establish a pattern and direction.

Usually a pattern will develop which angles towards a new high point. It may point directly towards this high point, but if two ground slopes intersect, or if one has become predominant, the pattern may trend downhill in both directions. When such a pattern has been developed the surrounding terrain in the direction it slopes should be examined for a prospective new high point. Then a new search line is drawn from the high point to a point inside the pattern and a new search triangle drawn and the procedure started over.

It may be necessary to start several new search triangles during a lengthy search, especially if it is on gentle slopes to begin with.

Sooner or later one of two things will happen. Either gold will occur in the sampling device so often and in such quantity that the prospector will know he has found the original source or it will pinch out so that he knows he has gone too far and has to retrace his steps making a careful search for the outcrop. Outcrops are a complex subject which will be discussed a little later in this chapter. But before this, a few more hints on conducting the tracking search.

Tracking searches are conducted by several methods. The most obvious and one that is used at all times during any prospect is the visual inspection of the surface at all times. The prospector looks for rocks similar to the float or pieces of rock which might contain gold in its native state.

The soil and its coloring are also indications of mineral content. Generally speaking, any soil discoloration is influenced mostly by larger rocks near or immediately under the surface. Radical changes in color are always worth investigating. These color changes often follow the identifying features given for minerals. Red is related to mercury; blue and green to copper; yellow and brown to iron. In inspecting soils or rocks for color changes be certain not to be fooled by organic growths such as lichens which often resemble mineral discolorations.

Another area to watch is sudden or abrupt changes in vegetation. In some areas vegetation grows thickly along one type of geological formation and not at all on another. Such an abrupt change in the fauna might indicate a contact zone

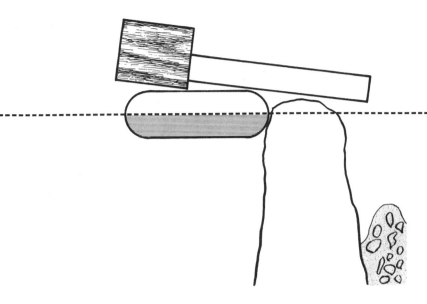

Using a rock, a dredge can be anchored in a position such as this.
However, watch the tailings pile and be sure to use a safety rope.

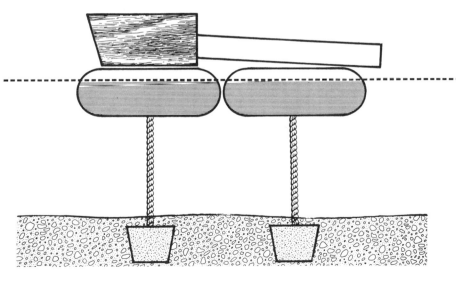

An anchorage such as this is always a last resort. The concrete anchors
are usually made on the spot, buried and left when the job is done.

between two geological formations—a good place for ores to be revealed.

One of the most important things to watch for, this statement also holds true when looking for a buried outcrop, is the holes of burrowing animals. Prairie dogs, gophers, badgers, woodchucks, ants and other creatures often burrow deeply and bring up samples of the earth several feet below. When conducting a tracking or lode search it is always wise to pan the material they have brought to the surface.

At all times during the prospecting trip the visual phase of the search is conducted whether the propsector is aware of it or not. His natural curiosity will lead him to investigate faults, dry lakes, hot springs, oozing of oily matters; black sands . . . nothing that is unusual should be left unstudied. And, if nothing else, the unusual should be examined only to eliminate it as a possibility of a clue to what the prospector is searching for.

One part of the visual search that bears serious consideration of the prospector is the discovery of "old diggings." These can be well defined as in the case of mines of only a hundred years ago which still have considerable visual evidence of their existence. On the other hand it make take a touch of genius and inspiration to identify the places where early natives or pioneers scratched out valuable surface deposits. Indians often scooped out only the enriched zone because they had no facilities for determining how the ores occurred. When they reached the vein they simply abandoned the digging or scratching out. These holes (not mines really) are now only the vaguest of depressions and sometimes only a lucky guess will reveal that a rich outcrop once existed.

Early Spanish miners of the western states were slightly more sophisticated and their diggings are more easily located. These mines have usually been worked by the more common types of placer and lode methods which do not require chemistry or powered machinery. As a result they may be beyond the capabilities for the advanced amateur who wants to do the same type of mining. However, they are always worth testing, if not sampling.

The mines of the late 1800's and early 1900's have pretty well been detailed in mining literature and if one is located in the natural process during a tracking search a little historical research will usually save a good deal of time in sampling. In the case of this type of discovery it often pays to examine closely the ore dumps when the evidence indicates the ore has lain idle for fifty to a hundred years. There have been many techniques of ore recovery developed over the last century and rapidly rising prices have recently taken place. Either of these might make the dump worth working. In addition, many minerals which were considered worthless a century ago are now much in demand.

Most of the preceeding information on the visual search is only "back of the mind" information that every prospector should have in case one or more of these isolated events occurs during his tracking search.

The tracking search is more scientific and occurs along the baseline. When the base line is lengthy sample holes are dug. There is no way to predetermine the

114

size or depth of the hole. Sometimes surface samples will suffice—other times a prospect hole several feet deep is dug. This must be determined on the site and is based on the prospectors estimate of the nature of the transporting agent and the depth at which the valuable mineral has been deposited.

When the base line is short, many prospectors resort to trenching the entire line (although this trench does not always follow a contour line, the best results are obtained when it is most nearly so.) This system is laborious at this stage of the game and the temptation is to take the samples at various points along the trench. This converts the trench system into a prospect hole system and defeats the entire purpose. To be indicative of the actual richness of a trench method, the entire gravel removed should be piled and sampled—since this procedure would defeat the pattern of accumulation theory it should rule out the trench method for the tracking search.

Trenching in the early stages of the base linessearch is probably no more revealing than the sample hole method and should be avoided until the last narrow lines. This method is far more useful in finding a buried outcrop and is most often used for this purpose.

The tools that are used for sampling on the search are those used for recovery of small gold placers. Usually the gold pan, dry washer and rocker are favored. Many old time miners always called these tools "samplers" and used them for that purpose only. The small portable dredge can rarely be used as once the search leaves the stream there is seldom enough water to run it. In addition, the basis of sampling calls for a very accurate estimate of the amount of gold being recovered and in many cases only a few flakes per pan will be found. With the pan, dry washer and rocker, the efficiency is much higher and the chances of fine gold escaping greatly reduced.

The final choices of tools will be determined largely by the amount of water available in the search area and the amount of gear the prospector can take. Many successful searches have been made with only a gold pan.

The proper method of digging a test hole is important. Since the biggest quantity of gold is probably at or near bedrock the prospector often resorts to digging a small hole to take his samples from considerable depth before washing them.

If the ground is not too well cemented, a small auger like those used in gardening will drill a four to six inch hole about four feet in a short time and bring up a sample from this depth. Another tool used for this purpose is a common post hole digger although it is more work. Even more dirt must be removed with an ordinary spade but it is better than the last resort, a pick and miners shovel.

To eliminate work and make it possible to examine as many prospect holes as possible the sample hole is made as small as possible. The gravel to be taken for testing is removed from the deepest portion of the hole, being careful not to let dirt removed from the hole fall back in as it will make finding colors more difficult.

If the tracking search is being carried on in the desert there are a few instances when the general rules can be broken or at least bent a little. If the base line crosses a dry wash or gulley, look for natural riffles just as though prospecting a stream. Try to imagine the wash with running water and look at such things as rocks, old car bodies, logs, anything which could create an obstacle to trap gold coming down the wash during flood times. If such an obstacle is off the base line in either direction it is a better place to sink a prospect hole than on the base line.

It is of little value to try to figure out where eddies were located during flood time as there is seldom enough time for them to deposit enough gold to be preserved.

A better bet in desert areas is where there is blow sand which may accumulate in dunes but more often the wind builds a miniature dune around a plant such as tumbleweed. These "miniature dunes" often have streaks of black sand on the surface. These are almost always a small eolian placer and often have very fine gold trapped with the black sand. This is another exception and it would be better to sample this area than a regular spaced location on the base line.

Before getting into the final source search portion of the prospecting trip, there are two other types of searches which are often used by experienced prospectors when the systematic procedures of the triangle type tracking search are not necessary. Both are much quicker and simpler and many prefer to use them falling back on a triangle search if these fail, or using the triangle search only in the latter portion of the prospecting search when the direction of the high point needs to be well defined.

The first of these searches could be called the "search line" method. In this system, the prospector merely fixes his sights on the most likely high spot and proceeds directly to it with only cursory investigation along the way. Prospects are examined at spaced intervals along the search line and the prospector never varies from the search line unless promising outcrops turn up which need investigation. This type of search is almost always combined with the "random search."

Random searches are those in which the prospector more or less follows a line towards high ground but has no fixed base line to explore. He may wander over the area in many directions exploring outcrops, contacts, faults, unusual vegetation, out of the ordinary eroded hills, or anything else that interests his curiosity.

The triangular search with its systematic exploration of baselines is by far the best for any search for a lode and it is not to downgrade it to say that the straight line and random methods are more frequently used because they are far quicker and often get the same results. They have a couple of defects which cannot often be overcome by the newcomer.

116

First of all, the prospector on a straight line or random search is often conducting a triangular search with more science than he cares to admit. Through long experience he can eliminate, or nearly eliminate, many of the marginal places where gold might not occur. "Horse sense" like this comes only after years of prospecting and there is no guarantee it will work in a new surrounding where the geology might be different.

The second deficiency also relates to experience and it is the ability to read a watershed with precision. Some old timers can read the slopes with insight that less experienced prospectors will not attain for years. In following a straight line they can almost always come up with the right decision when it is time to change a high point from just testing one or two prospect holes.

Most of these two individualistic types of searches have been made when the search line followed a drainge gulley or dry wash. In this case either of the searches is more practical than a triangular search. Fortunately almost all lode prospecting trips will follow such a watercourse until the last few miles. Then the triangular search will help establish the most likely high point.

The first rule for a lode search would be to follow all watercourses whether they have running water or not until the depression degenerates into a watershed that might have drained two or more high points, then conduct a triangular search on the "least work" basis until a pattern is developed that indicates the gold has come from the smallest area of suspect high ground possible. This will be the high point of the tracking search and the most likely spot for the outcrop to be hidden.

Once the tracking search has been completed and the prospector is reasonably sure he is at the general location of the lode, the third phase of the search begins—hunting for the outcrop or source of the float. Once again it is necessary to define outcrop as the miner knows it and not as the movie maker defines it. The term is greatly misunderstood by the general public. People expect to find a huge mass of white quartz speckeled with gold which juts upward 20 to 40 feet. Although this is sometimes the case (minus the gold almost always); the term outcrop has a very specialized meaning to miners.

Outcrop simply means the place where a mineral deposit comes in contact with the surface of the earth. In some cases it does protrude above the surface many feet, but more often it has been eroded level with the surface and sometimes is buried beneath several feet of overburden.

When the prospector finally arrives at his original site of the gold, there may be no indication at all an outcrop was once standing. In fact, buried outcrops are probably the best single prospect for an individual to locate a bonanza for most of the standing ones have already been investigated.

If the outcrop is visible, the prospector has no further search problems and can proceed immediately to testing and investigating the outcrop. Sometimes an outcrop will be level with the surface and can be located visually or with only the least amount of scratching in the dirt. Generally, the contact spot will be buried from several inches to many feet and the search for it must be more scientific than either the straight line or random search and probably more systematic than the triangle type tracking search.

First examine the surrounding terrain. It frequently occurs that the high spot of the tracking search is not the highest spot around. This location may be on the side of a mountain, in a gap between two mountains, a hollow on the side of a slope or in any one of a dozen other spots well within, but not at the head of a drainage system.

By using a little basic geology it is usually possible to determine which direction the soil has been carried and is currently due to erosion. This too, should be in a downward slope.

Down hill, from the suspected location of the outcrop, the search begins. The base line is drawn across the slope at right angles to the grade. The base line is usually informal and testing begins with the simpler methods. The starting point is about the center and work progresses to each side alternately to eliminate working the wrong side of the line first.

Generally speaking the prospector will use the digging tools he used most on the tracking search. These include the auger, spade and posthole digger. If none of these work, or if the pick and shovel is indicated due to the resistance of the earth, trenching is more practical and usually faster.

This system was introduced in this country by Cornish miners and is called "costeaning." The general practice is to dig a narrow trench at right angles to the down slope movement of the heavy mineral. The trench is begun at the most likely spot and dug several feet in each direction. Sampling is done every few feet at predetermined spots (often about five feet apart). As soon as the colors are picked up in good detail, the hole or trench is deepened at this spot to try to locate the outcrop.

If there is no evidence of a hardrock lode discovered in the search digging, and if the material is fairly well cemented but still turns up a reasonable amount of color, there are two distinct possibilities.

First, it may be that this is a residual or alluvial placer which has been buried under the overburden. If it is the former, it is probable that it is the residual of a leached out outcrop and may be all that remains. If this is the case it may be very rich and suitable for working by hand. When the hardrock portion of the vein is reached the rock may have too little gold in it to support any type of commercial mining and be too expensive for the amateur miner to get out.

If the discovery is a buried alluvial placer, it would seem to indicate that further up hill there would be a residual or at least a rich outcrop buried perhaps even deeper. Without extensive test drilling or a geologists report it might be impossible to find the original source of the gold. In fact, it may no longer exist, having been eroded away eons ago.

There is always the possibility (especially in California) that a hidden placer located in a lode search is the channel of an ancient river. Such a deposit can be quite extensive and economically feasible to promote. There are many such deposits in the West which can no longer be worked due to the restrictions on hydraulicking. However, a rich section of such a placer might be located during a lode prospecting search which would lend itself to hand methods.

Once the prospector has located an outcrop and determined that it is an outcrop and not a buried placer of some type, the time to sample the outcrop has come. This is not the type of sampling that could be called gathering material for an assay. Rather this is a type of testing that consists of getting enough representative rock to discover if the outcrop has any gold or other valuable minerals to warrant the trouble and expense of an assay and test drilling. Often this test is done in the field.

Naturally, all the mineral identification methods elsewhere in this book, or in a textbook on mineral identification, that are within the abilities and equipment of the prospector can be tried. The trick, like it is in more extensive sampling, is not to fool yourself. After discovering an outcrop the lode prospector is the most likely person to become a highgrader—and he always highgrades himself into believing the outcrop is richer than it is.

Yet, a good deal of suspicion of values should always be in the mind of anyone investigating an outcrop. Many are pure white quartz and valueless. However, a few pieces should be broken off with the rock hammer and looked at under a magnifying glass of at least ten power. If this examination shows any sign of gold or other minerals, several pieces should be broken off and processed with the mortar and pestle system given in the sampling chapter.

If the outcrop is of a striking different color, or if any of the other tests for valuable minerals hold true, the prospector should proceed to sampling and procure a scientific quantity to be assayed in a laboratory.

However, the only clue to added richness may be color changes in the surface rock. These stains usually come from water and many years of chemical action. The colors are seldom striking but they often mean valuable minerals beneath the surface. In the case of an outcrop meeting, or just below, the surface they often indicate an enriched zone of leached out (decomposed by chemical action) ore that can be fabulously rich. True, an enriched zone is rarely, if ever, of any great extent, but the ore found here often runs to the thousands of dollars per ton in richness. Any outcrop should be investigated to see if it has an enriched zone which the prospector can clean out by hand. For even though small, these finds often turn out to be bonanzas.

14

SAMPLING THEORY

Sampling is one of the easiest, yet most complex, problems in mining. Easy because the simplicity of gathering material does not involve the physical labor required in steady mining. Complex because there are many rules which if not followed can result in salting (an over rich sample) or leaning (a sample with too much country rock). If the sample is salted, much time and money can be spent trying to get out values that don't exist in the quantity the sample showed. If it is leaned a valuable lode or outcrop might be abandoned because it is apparently barren.

The prospector who searches for the sheer pleasure of finding valuable minerals will seldom be involved in any sampling to determine whether a deposit is commercially exploitable. Most of the information in this chapter will concern sampling outcrops and placers only to determine if they are worth working on a limited basis to recover a reasonable amount of precious minerals in proportion to the time spent in mining. Needless to say, if the samples prove to be very rich a geologist or mining engineer should be consulted—after the property is claimed or legally tied up with a lease so that any profits will accrue to the prospector.

Sampling is simply the process of securing representative quantities of ore so that it can be tested to give an indication of the richness of the whole placer or lode. The two major problems of sampling are: 1. Gathering the ore from equally mineralized locations, and 2. Mixing and dividing the ore so that a representative sample is obtained. Perfection in sampling is impossible so all sampling is a compromise. Following the rules will eliminate the possibility of a huge error and this is the overall purpose of sampling.

The first step is the gathering of the material. Sophisticated methods like taking a portion of the daily production are usually out of the question because the amateur prospector is usually testing, not mining.

If the material of the deposit is lightly cemented the prospect hole method can be used. Augers, spades or posthole diggers are tried first and if the soil resists, the pick and miner's shovel.

These methods are practical only where the sample is being taken from a deposit relatively close to the surface. The depth of the hole is usually only about 24 to 48 inches and this is not enough depth for placers which are about the only type of deposit for which the prospect type of hole is adequate. Generally speaking, if the post hole method samples show any colors in quantity, deeper drilling is used as a secondary sample.

If enough holes are dug in proportion to the size of the placer and spaced so that they cover all types of material in the deposit, the sample will give a fairly good indication of the richness of the deposit to the depth sampled.

In some cases, especially if a back hoe is available, the prospector will trench an area. This system gives an even better sample material than the pit method but due to the greater quantity of material removed the problems of contamination and reducing the sample to a workable size are greater.

Pits and trenches are usually satisfactory to determine if a surface placer is worth working. Since this is usually all that really interests the amateur prospector the search stops here and he begins his recovery operation. If more information about the size and richness of the entire deposit is desired the next step is drilling. Core drilling is beyond the realm of most amateurs and the equipment is expensive. However, if they do have the deposit core drilled the general idea is to get to bedrock if the placer is an ancient waterway; or, to drill through to the no-value area if it is an alluvial placer. The sample drill holes should be spaced about like the original ones but not necessarily in the same spot. They should be drilled until bedrock is indicated, the depth noted and the sample taken from that area. The biggest danger in test drilling placers, especially by hand, is that the drill sometimes encounters a large boulder which gives an indication of false bedrock. If several holes are drilled the occasional boulder will reveal itself by stopping the drill far short of its normal depth.

Occassionally the drill should be allowed to penetrate the indicated bedrock to see if the original stream bed had a false bedrock. Many very rich deposits have been found by this method.

Most amateurs will never get to the point of trenching or core drilling but most will come across the face of an ore body or a promising outcrop they want to sample. This can run into a lot of work sometimes but these problems can be solved.

First examine the deposit and determine how much of the vein or outcrop needs to be sampled. Remember, that in most cases the amateur prospector is only interested in the vein. It is not usual for him to open an adit and be concerned with the removal of a lot of country rock. It is more likely that the average amateur will just clean out a vein to a shallow depth and therefore if the vein is narrow, his sample can be highgraded from the vein itself. This type of highgrading a sample is all right but, if the deposit involves serious mining, a fair proportion of the country rock must be mixed with the sample to get an indication of how much the overall mineralization ratio is. If the vein is extensive, it is a good idea to take several samples along the face of it as color is rarely an indication of some types of mineralization.

Taking samples from rock will vary with the nature of the rock itself. If the vein material shatters or breaks easily samples can usually be obtained with nothing more than a rock hammer. As the rocks become more resistant to this method a small drill, chisels and pry bars can be tried. As a last resort, a small hole can be drilled and the rock shattered by blasting.

Besides placers and lodes, old mine dumps are good prospects for sampling. Modern technology and testing procedures may reveal something the original miners overlooked. Mine dumps can be sampled easily by the pit method. Due to the makeup of most dumps the material will not support the sides of a hole at much depth and it is necessary to timber a prospect hole or to run some type of casing.

Once the samples have been taken it must be gathered in one place with the least amount of contamination with surface material as possible. Usually it is gathered on a large piece of canvas. A large piece of metal would be ideal but these are seldom around when it comes time to divide a sample.

When the sample is taken from a relatively small area as the face of a vein, the canvas can be brought near the work and the sample placed directly on it.

Once the sample material has been gathered it should be reduced to as even a size as possible. In the case of placer gravel being tested for gold values only further reduction in size is seldom required and the sample is divided and panned. If the sample has been obtained by cutting it out of rock it is usually necessary to break the rock into smaller sizes before dividing. This is done on the premise that if a very rich piece of large ore finds its way into the final sample the size is of importance in giving a false reading. For instance, a piece of quartz that is several inches in diameter and speckled with 40% of natural gold could readily give a false value if it survived the dividing procedure to get into a one pound assay sample.

For field testing or for an assay the material must be further crushed before it is divided into sampling size unless it is composed entirely of material that is less than one inch in diameter. One of the best ways to break the rock down is with a maul or small sledge hammer. So that pieces do not fly away when they are broken, an iron ring is often used to contain the samples. These can usually be found in metal supply shops or junk yards.

This procedure should not be confused with the later procedure of reducing the material to a powder with a mortar and pestle. Right now, before division, the idea is to reduce the ore so that the largest piece in the sample is in small enough proportion to the entire test material gathered that if it makes its way to the final sample it will have little effect on the end result. As a rule of thumb no piece should be bigger than an inch in diameter.

Once the material has been reduced to the required size it is piled on the canvas or tarpaulin for division. There are two ways to do this depending on how much material is being divided. Either is a good method but do not change from one to the other in the middle of the division.

Coning is the most often used. First the sample is worked up into a cone in a random manner. Naturally some of the larger pieces will be near the bottom but this cannot be avoided and does not make much difference in the end result. Once a more or less symmetrical cone has been made the prospector flattens it with the back of his shovel. This makes a disc which is about a tenth as high as it is in diameter. In other words, if the diameter of the disc is 60 inches, the height should be about six inches.

This disc should be as close to a circle as possible and it is quartered by

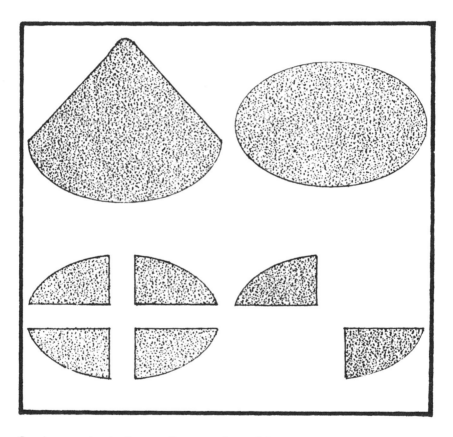

Coning method of sampling consists of four steps: Make a
cone, then flatten it. Next divide it into four equal portions.
Then discard two of the sections and start over. Repeat until
the sample is of the size required for an assay. Entire procedure, plus a method for doing the same with a blanket,
is described in the text.

drawing two lines with the shovel that cut the center and are perpendicular to each other. Quarters on the opposite sides are removed from the pile and discarded well off the canvas. Then the pile is built back into a cone, flattened and quartered again. The procedure is repeated until the weight of the remaining two quarters equals the weight of the desired amount needed for assaying.

The rolling method works well if the material is to be divided is dry or very nearly dry. The amount that can be mixed this way depends on the quantity of the sample, the moisture content and the size of the canvas. If all these factors are favorable it is the best method of mixing.

The material is placed as near one corner of the tarpaulin as possible and this corner is pulled across to the opposite corner so the tarpaulin forms a triangle. The canvas is then laid flat and the opposite corners pulled together. This procedure is repeated until the material is well mixed. It is usually necessary to pull the sheet a little more than half way to insure proper rolling and mixing.

Once the material is mixed well, all four corners are pulled together so that the material is gathered together in the center. The material is then flattened as in the conning operation and then quartered with the opposite ends being discarded. The process is repeated until the sample is reduced to the desired size.

After the sample has been reduced to a manageable size it is placed in a plastic container that insures it will not become contaminated. At this stage of the procedure perhaps as much as 1000 pounds have been reduced to only a few pounds and even a small quantity of surface soil can change the final test results considerably.

Preparation for assaying of the sample can be done by the prospector and in some cases he can do his own testing. This is especially true if the sample is to be tested only to see if gold or some other valuable mineral is present without determining how much is present as a ratio of the entire amount. The following suggestions also apply when a piece of rich float is to be tested to determine whether it is worth following in a tracking search.

Before any serious testing can be done it is necessary to reduce the sample to the finest grit possible. To do this the sample ore is placed in small quantities in a mortar where it will be ground with the pestle. If the ore pieces are large enough in diameter that a hammering action with the pestle is necessary to break it down, a shroud should be constructed out of canvas so that small pieces will not fly away when they are shattered. This can usually be done simply by wrapping a piece of cloth around the handle of the pestle and letting it drape around the mortar.

The sample should be ground to a powder between 30 and 50 mesh. If the prospector wants to be technical he can obtain a geological screen and sift the ore and re-grind until it is all of the same consistency. But for the usual run-of-the mill test the ore can be considered ready for assay or gest when it reaches the consistency of a fine powder.

It is a good idea to gather all of the material ground in the mortar in a plastic bag which can be sealed against moisture and air. There are many of these for sale in the household section of supermarkets and any, except the lightest, will work.

If the sample is to be tested in a professional assay office the bag with its ground sample material is delivered to the assayer. If the prospector is testing only for heavy minerals such as gold, the procedures will be found in the final concentration chapter of this book.

15

OTHER VALUABLE MINERALS

ALUMINUM: Aluminum is never found as a native metal. Its principal ore is bauxite. This mineral has been found in Alabama, Arkansas and Georgia in this country. Over the world it occurs mostly in the tropics, or in areas that were tropical in ancient times. It is usually found in a weathered surface deposit. There are several minerals in the bauxite group and they vary in color from white to dark red-brown. It is frequently found in matrix which resembles a hard brown clay. Luster, dull; hardness, 1-3; specific gravity, 2-2.5.

ANTIMONY: Occasionally, antimony is found in its native state in the form of thin plates. Actually, it is exceedingly unlikely that a large deposit would be found but a few grains might turn up in a dredge or gold pan. In its native state, antimony has a tin-white color, a hardness of 3-3.5 and a specific gravity of 6.6-6.7. It is associated with stibnite, galena, barite and cinnabar. The first of these, stibnite, is the principal ore of antimony. This mineral has been found in many states with the finest crystals coming from the mines at Manhattan, Nevada. It is often found in quartz where low temperature veins were formed. It is crystalline and is peculiar in that the crystals are sometimes bent. In nature, stibnite has a steel-gray color. Luster, metallic; hardness, 2; specific gravity, 4.5-4.6.

ARSENIC: Often mistaken for native antimony. Many of the tests are the same and it is associated with the same minerals. Color is tin-white and its crystal structure is similar to antimony. One way to tell the difference between the two minerals is to heat them. In a candle flame antimony will form a metallic globule, arsenic will not; both give off a dense white smoke but the arsenic will smell like garlic, antimony is almost odorless. *Arsenic smoke is poisonous* and candle tests should be avoided by the amateur. If either mineral is suspected, it should be assayed by a professional. Some large masses have been located in Arizona. Luster, metallic; hardness, 3.5; specific gravity, 5.7

BARITE: A rather common mineral across the United States with considerable quantities often found in the middlewest. In western states it is often associated with ores of copper, lead and silver. Usually crystalline in form (often huge) it is also found in masses. Barite is found in many colors ranging from colorless to: blue, yellow, brown, red and hues of these colors. Color usually means nothing to the commercial user, but rock collectors place different values on them. Luster, glassy; hardness, 3-3.5; specific gravity, 4.2-4.6.

BERYLLIUM: This is a very brittle metallic element which is used mostly in the electronic industry. Its principal ore is beryl. In transparent form, beryl is a gem. It is often mistaken for quartz which it resembles. Beryl occurs in crystal (often gem quality) and in columnar masses. Colors range through white, blue, green, yellow and pink. Beryl is almost always found in pegmatite material and is seldom found in a high temperature vein. Luster, glassy; hardness, 8; specific gravity, 2.6-2.8.

COAL: Coal is a fossil fuel and not identified as a mineral. In recent years its economic importance has greatly increased and finding, if not immediately developing, coal deposits can be quite rewarding. It is easily recognized and identified in the field. In color it is black and burning a small piece produces the unmistakable odor of coal smoke. Large deposits have been found in almost every state but they are seldom economically feasible to work.

CHROMIUM: A valuable metal both for plating and as a metal. Its only known ore is the mineral, chromite. It has been found in New York, Maryland, North Carolina and California. It is crystalline in structure and sometimes slightly magnetic. In this country it is generally found associated with serpentine but these deposits are generally small. It also occurs in altered basic rocks. It has been found as placer deposits in serpentine areas. Good chromite crystal specimens are extremely rare and very much in demand by collectors. Luster, resinous; hardness, 5.5; specific gravity, 4.6.

COPPER: Copper is an element which is often found in its native state. This is occasionally in crystal form (high value to collectors) or in masses without recognizable crystal forms. Most native forms of copper are found in the midwest in ancient lava forms. The only commercial deposits of this nature were found in northern Michigan. The glaciers carried several small masses southward and these nuggets still occasionally turn up today, although most were probably found by Indians who made jewelry of them hundreds of years ago. Easy to recognize by its copper color and its malleability, it can be flattened by pounding. Color, copper red; luster, metallic; hardness, 2.5-3; specific gravity, 8.9.

COPPER ORES: Generally a commercially valuable deposit of copper is found as a sulphide. These minerals would include chalcopyrite, chalococite, bornite, tetrahedrite, cuprite, malachite, chrysocolla and azurite. These are easily recognized by their colors which are usually green or blue although some are brown. Generally speaking, the color is striking enough to call attention to a copper ore and many of them are prized as mineral specimens or gem cutting

126

material. For more precise identification consult a mineral identification table for each mineral.

GEM MINERALS: Almost any of the minerals can be classed as gem materials for the rockhound or cutter. A list of the common and uncommon gems found in the United States would include almost every gem known to man. Here are just a few to show their widespread occurrence: Diamonds in California placer fields; diamonds in Arkansas; sapphires in Montana; amethyst in Arizona; opal in Nevada; and turquoise all over the West. And, Eastern states have their share of gems, too. Gems are most easily recognized for their beauty and colorful stones, either transparent or transluscent, should be suspect. Since many of the gems are heavier than sand, the gold panner or sluicer often comes across them in his concentrations. The early California miners often threw away diamonds because they didn't know what they were. The same is true of the gold miners in Montana who threw away sapphires and rubies from their concentrations only to learn later that there were valuable commercial deposits nearby.

GOLD: The most sought after element by weekend prospectors, it is easier to recognize than most believe. It is most often confused with iron pyrites or mica which have similar colors. Iron pyrites can be eliminated by their hardness, about 6. Mica is about the same in hardness as gold, but under the pressure of a flat piece of metal, gold will flatten—mica pulverizes. Gold will scratch, but mica leaves a trail of dust. All gold is yellow in color but is usually found with some silver, the more silver the lighter the yellow color. Pure gold is soluble in aqua regia, but remember, gold which has a high silver content is soluble in other acids. One of the best checks of gold is its specific gravity. After once seeing native gold there is little chance that even the beginner will mistake it. Hardness, 2.5-3; luster, metallic; specific gravity, 19.3.

GRAPHITE: Most often graphite is found in this country as isolated crystals in schist or marble. These are seldom commercial deposits although they might have some value to rock collectors. A large deposit would be almost an instant commercial success as most of our graphite is imported from Ceylon and Madagascar. Since most graphite crystals are so small they can only be seen clearly under the glass, most of the identification tests are made in a laboratory. Large masses can be simply tested. Its native color is black when scratched on paper it leaves a pencil type mark. It is soft enough to be scratched with a fingernail. Luster, metallic; hardness, 1-2; specific gravity, 2.3

GYPSUM: It is highly unlikely that any large deposit of gypsum is still unclaimed in this country. These are very profitable and large companies have searched the country for them. Commercial deposits almost always occur as massive beds of sedimentary rock but can also occur as crystals in clay or crystallized cavities in limestone. It is possible that some of these latter have been overlooked and they might be of value to rock collectors. In nature gypsum almost always occurs as white, but can also be brown, yellow or red. It is extremely soft and can be easily scratched with a fingernail. Luster, silky (sometimes glassy); hardness, 2; specific gravity, 2.3.

IRON: Native iron rarely is found because it oxidizes so easily. All commercial deposits are from ores of iron, the principal one being hematite. Ores of iron are usually very easy to recognize because of their brown color and heavy specific gravity. Some, with a high percentage of iron, will give off a familiar ring when banged together. Native iron is found so seldom that it has a high value to rock collectors. It is steel gray in color and crystalline. The most common occurrence is in meteorites but it is sometimes found in basalt. Luster, metallic; hardness, 4.5; specific gravity, 7.3-7.8.

LEAD: Galena is the principal ore of lead and it is often rich in silver, too. Galena occurs in crystal form and large crystals are frequent. Sometimes it occurs in minute crystals which appear to be large masses. Its color is lead gray running to a silvery color and among prospectors the belief is strong that the lighter colors contain more silver. Only an assay will tell. Galena is easily identified in the field by hardness, it will just barely scratch with the fingernail; held in the flame of a candle, it will produce a globule of lead. Luster, metallic; hardness, 2.5; specific gravity, 7.4-7.6.

LIMESTONE: Limestone deposits must be large to be commercially exploitable and most of these have been found. It has little value to the rock collector. In nature, limestone is grayish and powders when scratched. Luster, vitreous; hardness, 3; specific gravity, 2.7.

MAGNESIUM: Magnesium is never found as a metal and is most often produced from chloride salt beds. However, the mineral, magnesite is an ore of magnesium and a large deposit might be valuable. Magnesite is usually found with serpentine or in sedimentary rocks. It is crystalline in structure and large crystals have been found. Usually it occurs in microscopic crystals forming a mass. In nature it is generally white, gray or colorless. Luster, glassy; hardness, 3.5-5; specific gravity, 3.

MANGANESE: Manganese is hard brittle metal that is rarely used as a metal. In modern times it is most important as an alloy. It never exists in nature as a native metal. There are no very large deposits of its ores but many have been exploitable. The principal ores are pyrolusite, psilomelane, rhodoschrosite and rhodonite. Pyrolusite is the most important ore of manganese and is extremely widespread. It is soft enough to scratch with the fingernail and black in color. Handling it leaves a sooty black residue on the fingers. Luster, metallic; hardness, 2; specific gravity, 4.7. Psilomelane is never found as a crystal but in masses. It is often associated with pyrolusite being covered with a sooty coating of this mineral. It is also black in nature but lighter shades have been found. It is easily distinguished by its hardness. Psilomelane should be assayed because samples which appear to be rich in manganese often have only a small percentage of the metal. Luster, metallic; 5-6; specific gravity, 3.3-4.7. Rhodoschrosite is a minor ore of manganese, but has some value to rock collectors. It is usually associated with copper, lead or silver ores but does occur in pegmatite bodies. It is easily recognized by its pink color but specimens of gray or brown have been found. Luster, viterous; hardness, 3.5-4; specific gravity, 3.4-3.7. Rhodonite is a pink mineral often mistaken for rhodochrosite but can generally be identified by its hardness, an absolute test is chemical and heat. This mineral is highly valued by rock collectors. Luster, glassy; hardness, 5.5-6; specific gravity, 3.4-3.7.

MERCURY (QUICKSILVER): Mercury is occasionally found native in nature. It cannot be mistaken as it is the only metal which does not solidify until the temperature reaches a minus 40 degrees. It is also characterized by its high specific gravity of 13.6. If a small globule of mercury is located in nature it probably indicates the nearby presence of cinnibar, the principal ore of mercury. It should be mentioned that mercury is often found in streams which are being placered. Usually this means that someone has been using mercury in his riffles and it has escaped. Cinnibar is most easily recognized by its brick red color and if the surface is oxidized the mineral breaks easily exposing the color. Another test is to pulverize a specimen and rub a penny across the powder. The penny will pick up the mercury and turn a silvery color. Luster, adamantine; hardness, 2.5; specific gravity, 8.1.

MICA: Mica is a thorn in the side of prospectors. The uninformed will always mistake the tiny yellow flakes of one type laying in a stream bed for gold. Placer gold rarely lays on the surface to reflect the sun but a positive test is to crush the mica with a knife blade, it will pulverize into a fine white powder. Mica is actually a family of minerals and only two are important to prospectors. Muscovite is crystalline in structure and has excellent cleavage. It has many commercial uses and there are several deposits in the United States. To be valuable the crystals must split into sheets no smaller than 1½ by two inches. In nature the color varies from white to yellow generally, but specimens of light red and green have been found. Luster, glassy; hardness, 2-2.5; specific gravity, 2.8-3.1. The other commercially exploitable type of mica is called lepidolite which is an ore of lithium. It is a rare mineral in this country and invariably found in pegmatite. Lepidolite comes in many colors including lilac, gray, green and yellow and is often confused with muscovite. Therefore, an assay by a professional is almost a must. Luster, pearly-silky; hardness, 2-4; specific gravity, 2.8-3.3.

MOLYBDENUM: Molybdenum is a white metal much used in tool steel alloys. Its principal ores are molybdenite and wulfenite. Molybdenite is a very soft black mineral which occurs as crystals but is often found as flakes or plates. It is easily confused with graphite. A streak made by molybdenite on paper is much like a pencil mark but tends towards the greenish side. A positive test is chemical. Luster, metallic; hardness, 1-1.5; specific gravity, 4.6-4.8. Wulfenite is a minor ore of molybdenum and contains considerable lead. It is almost always found as crystals which are generally thin. Most locations have been found in desert areas where the climate is arid. In nature the colors are red, yellow, brown and gray. Luster adamantine; hardness, 2.7-3; specific gravity, 6.8.

NICKEL: Nickel has not been found in commercial quantities in this country. Almost all production is a byproduct of copper ore. The principal ores of nickel are: garierite, millerite, pentlandite and niccolite. They will not be discussed in length as most of them resemble other valuable minerals and have a high specific gravity. Any well informed prospector would bring them in for an assay just on general principles.

PLATINUM: Small amounts of platinum have occurred in placer deposits in the United States. It has never been found in lode and such a deposit would be very valuable since this country imports all platinum needed for industry. There are few other metals which can be confused with platinum. Its color, silvery, and its high specific gravity are usually enough identification. It is harder than gold but very malleable and can be pounded into sheets. It is most often found with gold in a sluicing or panning operation. It is sometimes found pure but most often it is alloyed with others of the platinum family: iridium, osmium, osmiridium, palladium, rhodium and rutheriom. This gives native platinum a wide range of densities which makes specific gravity difficult to determine. In its native state placer platinum can vary from 14 to 20 in specific gravity. Luster, metallic; hardness, 4-4.5; specific gravity, 21.5.

SILVER: Silver is another metal which occurs in nature in its native state. After almost two hundred years of prospecting, there is probably little native silver left. It might turn up in the sluice or pan and could be mistaken for platinum. Silver can be identified by its specific gravity and the fact that it is the only white metal that is malleable and which will dissolve in acid. Luster, metallic; hardness, 2.5-3; specific gravity, 10.5.

SILVER ORES: Most silver produced today is from four principal ores: Argentite, ceragyrite, bromyrite and pyrargyrite. Argentite is nicknamed "silver glance" by miners and is the most frequently found silver ore. It is crystalline in structure but usually found in huge masses little resembling crystals. It is dark lead gray in color and will tarnish to a dark black and discolors quartz associated with it. It is extremely rich in silver (about 87%) and under a blowpipe will fuse into a silver globule. Luster, metallic; hardness, 2-2.5; specific gravity, 7.3. Bromyrite and cerargyrite differ only in that bromyrite has about 25% to 40% bromine, and cerargyrite has about the same percentage of chlorine in their chemical makeup—the rest of their composition is silver. Miners have nicknamed both the minerals "horn silver" because it can be cut with a knife. In nature the color is from colorless to gray. After being exposed to light, horn silver will darken to a violet-brown color which also helps to identify it. These minerals will melt in a candle flame and produce a silver globule under the blowpipe. Luster, adamantine; hardness, 1-1.5; specific gravity, 5.5. Pyrargyrite is found in low termperature veins and often crystallized. It is deep red to nearly black and contains much antimony. Luster, adamantine; hardness, 2.5; specific gravity, 5.8-5.9.

SULPHUR: Sulphur is an element that is found in many minerals composing the group called the sulphides. It is also found in its native state and is commercially exploitable. Most of the economical deposits in this country have been found in Texas and Louisiana. In nature it is easily recognized by its yellow color. An interesting feature is its low melting point, just over 100 degrees. It will melt in a match flame and burn in a candle flame. The most distinguishing feature is the odor it gives off when burning which is an unmistakeable smell. Sulphur is often found in crystal form in isolated areas and these have a high value to rock collectors. Sulphur crystals should not be handled since the heat of the hand will cause them to expand and perhaps crack. Luster, resinous; hardness, 1.5-2; specific gravity, 2-2.1.

TALC: Talc is a mineral which is easily recognized in the field. Its color is generally white but it often has a greenish tinge and is sometimes green in color. Extremely soft, it can easily be scratched with a fingernail. It is crystalline in structure but is rarely found this way. Most deposits are massive. While most talc is ground into powder, finely grained masses are called soapstone and sold as blocks. Talc can be easily powdered in the field or carved with an ordinary pocket knife. Luster, greasy; hardness, 1-1.5; specific gravity, 2.7-2.8.

TIN: Tin is a very valuable metal which is never found native in nature. However, its principal ore contains almost 80 percent tin and is heavy enough to be panned or sluiced. There are no commercial deposits in the United States but it has been mined here on a small scale in recent times. Tin has been found in placer and in lode in almost every western state and many of the east. Alaska has several commercial deposits which have been exploited. The principal ore of tin is cassiterite which ranges in color from yellow to black and is occasionally found red or brown. It is crystalline in structure and is found in lode this way. In placer, cassiterite is found as water rounded pebbles. It is identified by its hardness, high specific gravity and non-magnetic properties. Luster, greasy; hardness, 6-7; specific gravity, 6.8-7.1.

TITANIUM: Titanium is never found in its natural state. There are two ores which interest prospectors and although both are commercially workable, the most important of the pair is ilmenite, which furnishes most of the world's supply of titanium. Ilmenite is found in many places in this country and has been discovered in the black sands of gold pans and sluices in the Pacific Coast areas. In its original state, the mineral is crystalline in structure, but it breaks down readily and finds its way to placer deposits. It is black in color and resembles many of the iron ores. An assay is necessary for identification and as the tests are quite complicated, they should be made by an expert. Luster, metallic; hardness, 5-6; specific gravity, 4.1-4.8. Rutile, the other important titanium ore, is also black and frequently found in crystal form. It has value both as an ore of titanium and to rock collectors. The crystals frequently form inside clear quartz crystals as replacement material. This is called rutilated quartz and has a value to collectors and gem cutters. Luster, metallic; hardness, 6-6.5; specific gravity, 4.2-4.3.

TUNGSTEN: Tungsten is not found as a native metal. Its principal ores are feberite, hubernite, scheelite and wolframite. Cuproscheelite and tungstite are minor ores. Ferberite, huebnerite and wolframite are so closely related chemically that most authorities describe them together in a group called either *tungstates* or *wolframates*. The differences are slight: feberite contains iron and tungsten; huebnerite contains tungsten and manganese; and wolframite contains tungsten, iron and manganese. This gives each a slightly different color. Feberite is black, huebnerite is reddish brown, while wolframite can be dark brown to almost black. There are many minerals which closely resemble these tungsten ores and a good assay is required for identification. Luster, metallic; hardness, 4-5.5; specific gravity, 7.1-7.5. Sceelite is very easy to prospect with a fluorescent light since it turns blue or yellow in shortwave ultraviolet rays. In nature it resembles quartz and many scheelite deposits have been discovered by using a fluorescent light on previously examined quartz surfaces. Its natural color is white although brown and green crystals have been found. Luster, adamantine; hardness, 4.5-5; specific gravity, 5.9-6.1.

URANIUM: So much has been written on prospecting for uranium since the boom of the fifties that little remains to be said. Government subsidies are no longer in effect and the ores no longer command the importance they once did although good deposits are still commercial prospects. There are some fifteen uranium ores but most writers only mention two. Serious prospecting for uranium means using electronic devices but the methods old time prospectors used are still useful. Uraninite is the principal ore of uranium and is commonly called pitchblende. It is crystalline in form but usually found in masses. It is black in color and somewhat resembles obsidian or volcanic glass. The difference can be quickly recognized by lifting a piece of uraninite, it is about five times as heavy as obsidian (specific gravity, 2). Luster, resinous; hardness, 5.5; specific gravity 8.7-10. The other better known ore is carnotite which contains both uranium and vanadium and is sought for both elements. During the big uranium rush it was discovered by geologists that what even the experts thought was carnotite was really something else. Only a specialized assay office can tell the difference. Carnotite is a canary yellow in color and is usually found in the cracks between rocks. Luster, earthy; hardness, 1; specific gravity, 4.1-4.5.

VANADIUM: Vanadium is not found as a native metal. It is derived from many ores, only a few of which will be described here. Most often it is found in minerals carrying some other valuable element. One of these, carnotite, is described in the paragraph on uranium. Roscoelite is a little known member of the mica group. It is green in color and sometimes mistaken for lepidolite. An assay for vanadium must be made for identification. Luster, pearly; hardness, 2-2.5; specific gravity, 2.8-3. Vanadinite is a lead and vanadium mineral generally found in desert areas. It is frequently found in large, well formed crystals which can be red, yellow or occasionally green. Luster, resinous; hardness, 2.7-3; specific gravity, 6.7-7.1. Descloizite and mottramite are closely associated minerals which are vanadates of lead, zinc, copper and vanadium. When more zinc is present it is descloizite; if copper is in larger proportion it is mottramite. It, too, is found in arid areas and usually as a secondary mineral to other ores. Luster, greasy; hardness, 3.5; specific gravity, 5.9-6.

ZINC: Zinc is never found as a native ore and is frequently found associated with lead deposits and occasionally with silver. There are several ores of zinc, four of which are important to prospectors. In geological and mining literature the word calamine often appears as an ore of zinc. The word was used by the English to refer to smithsonite and by American writers to refer to hemimorphite. And, in much of the literature it is used interchangeably. Hemimorphite is normally white but can be yellow, green, blue or brown. Sometimes it will fluoresce orange in ultraviolet light but this is not a reliable test. Luster, glassy; hardness, 4.5-5; specific gravity, 3.4-3.5; Smithsonite is a zinc carbonate which can be brown, green, yellow, gray or even white. It is sometimes mistaken for similarly colored quartz which it resembles but can be identified by its hardness. Luster, vitreous; hardness, 5; specific gravity, 4.3-4.4. Sphalerite is popularly known to miners as zinc blende and is the most valuable ore of zinc. It contains 67% of the metal as well as gallium, indium and cadmium, and is also the principal ore of these metals. It is often found in crystal form and is usually black in color. In appearance it resembles galena but a blowpipe test for galena will eliminate it. Luster, resinous; hardness, 3.5-4; specific gravity, 3.9-4.1. Hydrozincite is a secondary ore of zinc that occurs in arid climates. It has no crystal structure and will turn a brilliant blue under ultraviolet light. Luster, dull; hardness, 2.5-5; specific gravity, 3.6-3.8.

16

TOOLS OF THE PROSPECTOR

While many amateur prospectors have a truck full of gear, the equipment and tools you need to get started getting your share of nature's gold will consist of only a few basic items. A minimum list would include: a shovel, two picks (one large, one small), a magnifying glass, a pair of tweezers (to pick out small particles of gold), a gold pan and a small glass bottle to keep recovered placer gold in. Outside of the gold pan, almost everything else on this list might be found around the house and garage. So a minimum investment might be only a few dollars.

As an amateur becomes more proficient, or when he finds a good placer location he wants to clean up, he will want to add many items to his gear. Some will be to do the work faster, others to do it more efficiently. Still others will be used to recover gold in places that take a special technique such as dry washing where no water is available. Almost anything an amateur prospector might want is available commercially. Unfortunately some of these items are quite expensive and it would not be practical to purchase them for a small operation. For those who are handy with tools, I have included many drawings which while not drawn to any particular scale, will be quite helpful in constructing such equipment.

The tools and equipment described in this chapter will have some notable omissions (the portable dredge, for instance), which are described in other chapters. Most of this equipment is very old and has been used for hundreds of years; other items, like the plastic gold pan, are very new and are just now winning wide acceptance among prospectors of all types. One thing should be remembered . . . ingenuity is the secret of success in recovering gold or any other mineral. It seems that a person always comes upon a good stream when he doesn't have the right equipment. Once a friend and I came upon several persons panning on a stream in the desert which seldom had water. We had no pans but were able to scrounge up two huge garbage can lids. They were clumsy and inefficient but we did get some gold. By reading this chapter, the amateur prospector will learn many ways to recover gold with merely the natural materials on hand.

GOLD PANS AND THEIR COUSINS. The first tool that the novice must master is the gold pan. Not only will the pan be used to concentrate placer gold and nuggets, it is useful for recovering many other types of minerals and gems. Besides being an excellent testing device at placer deposits, the pan is indispensible in most types of lode prospecting.

There are four materials commonly used to make gold pans: steel, aluminum, copper and plastic. The most familiar is the steel pan but many prospectors favor the aluminum pan since it does not rust out. In the past ten years there has been a strong trend to the plastic pan which is not only indestructible but also has several unique advantages. It may soon replace both the steel and aluminum pan. The copper pan is used in amalgamating final concentrations and will be discussed in the section on final concentration methods.

A good gold pan should be of heavy gauge construction and have a beaded rim to give it strength. Pans come in sizes of four inches to 24 inches in diameter. Most amateur prospectors have at least two different sizes and several sizes are not unusual. Each has its advantage. The size of the pan depends upon a lot of things, the speed of the water; the fineness of the sand, the amount of gold, etc. My favorite sizes are 18 and 24 inches, but this varies with the individual. When learning to pan it is a good idea to start with a smaller pan until you become proficient. The sizes below eight inches have a special use and will be discussed separately. For those who have never panned gold, the 12 inch size is a good beginning. With this you can learn the basic principles and graduate later to larger pans.

Many people have little faith in the gold pan. This is something that has not changed in the last 2,000 years. An ancient Greek, Theophrastus, wrote in 300 B.C., " . . . in this work (panning) there is much art to be used; for from an equal quantity of sand some will make a large quantity of the powder (cinnabar), and others very little, or none at all." Don't let Theo throw you, panning is truly an art, for there are those who recover far more than others. Yet, I have yet to meet any person, young or old, who could not master the techniques of panning. It is simplicity itself to recover 75% to 90% of the gold in a pan. Where people go wrong is in trying to write down the instructions for panning.

Written instructions for gold panning are complex and sound much more complicated than they really are. The best method is to simply watch an experienced gold panner. Sometimes this is not possible and the procedure can be learned equally well in your own back yard if you have some sand and water. And, even if you live in an area which is not known for placer gold, virtually any stream can be used to learn the basics of panning gold. Almost all western streams have sand, and this sand usually contains some black sand which is a form of iron that is much heavier than regular stream sand. By learning how to concentrate the black sand you will have mastered the identical principles of concentrating gold and introduced yourself to something that will be quite a nuisance when you go after gold—this same black sand.

If you happen to live in one of those places where there are no streams with black sand, or if you have no streams near you, there is another method for learning to pan gold. Simply mix small pieces of lead (the smallest of shot to begin with, and lead filings later to perfect your techniques) into common sand. The lead is heavier than black sand and quite easy to concentrate in the pan.

The secret of recovering gold is a well known physical fact. Simply, that gold is heavier than sand. Gold has a specific gravity of 15 to 19.3 compared to the specific gravity of common sand of about an average two or three. Black sand usually averages about five to seven, while lead's specific gravity is 11.

All through the recovery of gold, placer especially, but including most lode deposits, this fact is empirical. Gold is heavier than any other mineral commonly found. By agitation, suspension, or other disturbing outside effects, the gold is allowed to settle first into a concentrated accumulation where it can be effectively recovered by simple separation methods. This rule will pop up again and again in this book: *That which is heavier sinks first and furthest.* It is the cardinal rule of all gold prospecting.

Gold panning uses agitation and suspension of lighter material to accomplish this recovery. More simply stated: The object of gold panning is to concentrate heavier materials (black sands and gold) by washing away the lighter materials by swirling the pan, bouncing it up and down, or stirring the material with your fingers. The heavier sands sink to the bottom and become concentrated. Extraneous materials such as silt, mud, sticks, pebbles, sand, etc., are either picked out by hand or washed over the side of the pan.

It is best if your original attempts at panning are practiced at a stream with running water. If no stream is available, simply fill a large tub with water, put sand in the pan, stir in the lead particles completely and follow the outlined procedure.

Regardless of the size of your gold pan, fill it about three-fourths full of sand to be concentrated then submerge it in the water. Many times in desert streams the water will not be deep enough to submerge it entirely. This makes panning difficult, but water can be dipped into the pan and the operation carried out this way.

Gold and other heavier minerals can most efficiently be concentrated when the material in the pan is nearly all the same size whether it be sand, clay or just plain dirt. The first object of panning is to accomplish this. Examine the pan of material and remove all visible large stones, sticks or other extraneous material. When you throw them away, be sure to get them well enough out of the way so they don't show up in your next pan.

Next submerge the pan under water and stir the sand with your fingers. If you are using river sand, several larger pebbles and rocks will come to the surface. These should also be removed and thrown away. If you do the stirring and sorting under water a lot of silt will come to the surface and wash away with the stream current. Be sure to observe this washing closely. If the current is too strong, a lot of your gold will also wash away before it has a chance to settle.

While you are removing these pebbles be sure that no particles of clay or hardened dirt are thrown away. These should be crumbled between your fingers and dropped back into the sand to be concentrated. Clay or hardened dirt clods often contain high concentrations of placer gold. When you are breaking up this material, it is best that the pan be out of the water as much of the material could be floated away by running water.

Once the bigger pieces of rock have been removed and all the bits of clay broken up, put your hand back into the pan and stir it up with your fingers. This will help release silt, bits of wood and other lighter materials that resisted the first procedure. This step may have to be repeated several times and will depend upon the nature of the material being panned.

The initial step, which requires many words to explain will actually require only a minute or so. The entire operation can be done under water if the current of the stream is not too fast. Stirring too violently will bring heavier materials to the top and let them escape before they can be concentrated. Where the pan cannot be immersed, the water must be dipped in and poured out. The same danger of letting heavier material escape is always present so be sure not to stir too violently.

While you are removing the larger rocks and agitating the sand with your hands, watch these pebbles with care. If there are any nuggets or small pieces of quartz with gold attached, then this is where they are most often found. Finding a nugget in the pan is a rare occurrence, but they do turn up now and then and a little scrutiny at this stage of the operation might bring a prize worth several days of work.

Almost everybody does the first procedure the same way with little variance between different gold panners. And those who proceed beyond this chapter to the more sophisticated methods of concentrating gold, will quickly note that every method from the sluice to the portable dredge, uses exactly the same procedures. Breaking, cleaning and sorting the sand is done by using grizzlies, parallel riffles, screens or fingers; just to name a few methods.

Once the foreign material has been removed from the sand to your satisfaction, the actual gold panning begins. The next procedure is difficult to explain partly because every person develops his own style, and partly because different soil materials require different methods. There are almost as many methods of gold panning as there are gold panners. But once the basic fundamentals have been mastered it makes little difference how you agitate your sand. The end results are the same. I will explain the more or less scientific methods and it should enlighten the serious amateur so that he can develop his own style of panning gold.

First, dip your pan back into the stream and raise it up full of water until it clears the stream. Now make a circular motion with the pan until the sand is well mixed with water. A quick jerk occasionally will help the sand loosen up and mix it with water. This has the effect of suspending all the sand in water and allowing the heavier materials to sink to the bottom. As soon as the sand is well mixed with water and moving well within the pan, tip the gold pan away from you and allow water and some of the lighter sand to escape over the side. Then dip the pan into the water and repeat the same operation. This is the actual "panning" and once the principle is understood, it takes only a few minutes to work even a large pan of sand. How many times the operation must be repeated depends on the consistency of the sand, the quantity of clay, and the proficiency of the operator. Some gold panners do the whole procedure under water, others do all their panning above water letting the extraneous sand escape as they pour water out. Which is best? I would not even hazard a guess, but I prefer to work under water if the current is not too fast. Whatever works out best for each individual is what he should use.

After several swirlings, dippings and pourings, you will begin to notice a small streak of black sand appearing around the outer edges of the sand. How much cannot be estimated. Sometimes it will be as much as a cupful; but more frequently, about enough to fill a tablespoon.

Once the black sand becomes visible to the naked eye, more care must be exercised in the panning operation to avoid loosing these heavier materials (which also contain the gold if there is any). Now the problem becomes to remove the lighter colored sand and retain the small quantity of black sand. One

137

of the best ways is to swirl the sand around in the bottom of the pan making a rough separation between the black sand and the white sand. Then separate it with your fingers and push the lighter material over the side. Be careful in this operation because much of the black sand will stick to your fingers. Wash your fingers off with water in the pan and it will naturally come to the bottom once more.

If the sand you are concentrating is particularly rich in gold, it will appear now as a part of the black sand. Actually, indications, or colors, should be relatively easy to spot although the values may be too small for the naked eye. An examination with a magnifying glass helps, but since the practice is to save all the black sand for later concentrations; this operation only gives confidence. If there happens to be a large enough quantity, the gold can be physically removed from the pan. This is done in the case of slightly large grains with a pair of tweezers. Most frequently, however, the gold is of such small size that it must be washed over the side.

As a note here I should mention that if you are using a plastic pan don't use the magnet for separation now. This can be used in final concentration, but the black sand has to be thoroughly dried. Wet sand will adhere to fine particles of the gold and these may be lost.

Washing the gold over the side may sound a little risky but a little practice will let you recover nearly all the gold. Use a wide mouth jar or similar glass container. (Don't use a tin can unless absolutely necessary as they often have seams that will trap small values.) Pour a little water in the pan and swirl the black sand and gold around until the gold appears as a tail on the trailing edge of the black sand. This is just another method of panning and since the gold is heavier, it travels slower around the bottom edge of the pan. When the difference between the gold and the sand is well defined, tilt the pan so the water runs off towards the center. Then place your thumb over the black sand and let the water come back to this corner of the pan. A little jostle and tilting of the pan will separate the gold which can then be poured over the edge with the water into the glass jar. Let the gold settle in here for the day and as the water from several pans starts filling the jar it can easily be poured off to make room for more with no loss of gold. Later, when the jar has a fair amount of gold it can be transferred to a smaller vial or bottle where it will probably be stored until displayed or sold.

After pouring the gold out you might notice that a few colors stick to the side of the pan and are intermingled with the black sand. These will be practically impossible to separate at this time by the method just described—besides the time it would require would be more than it takes to work another pan full of sand. (And don't forget, there are often colors too small to be detected with the naked eye that can be recovered later.)

Have another glass container (plastic works well too) and using the same method pour in the black sands. These are saved to be treated under the methods described in the chapter, "Final Concentration Methods." Notice that even now all of the black sand (and consequently some of the gold) cannot be poured out. *Do not place the pan back in the water to fill it with sand.* Rather, use a shovel to get fresh sand. This way the small amount you couldn't pour out will be concentrated in your next pan. Instead of losing the values from every pan, you will only lose them from one, the last pan full of the day.

One note on the steel gold pan. To make it easier to see the gold, they can be colored by heating them in a fire until they turn dark blue or black. Aluminum pans cannot be so treated and the plastic pans are already black.

Plastic pans are offered in smaller diameters but processing less sand per pan is often outweighed by other advantages. Compared to a similar pan made of steel, the weight is less than half. This is a considerable savings in work considering how many times the pan must be lifted each day. The material is usually ABS plastic that cannot be affected by weather, temperature change or any normal use. Unlike the metal pans, it does not dent when dropped.

The center, or drop, is much more abrupt than on the metal pans and this effectively retains the heavier material somewhat better. Additionally some plastic pans have sort of riffles which also help to separate heavier material from the bottom of the sand.

Plastic pans have virtually replaced the metal ones in many areas. They are easy to use except the black color of the pans tends to obscure black sands. However, this same background does make the gold stand out vividly.

MINER'S HORN. No doubt you have seen gold pans of six to eight inches in diameter and wondered what such a small pan was used for. It is wrong to imagine these are toys, they serve a very useful purpose. (Some of these are actually shaped like a horn and today they are often called backpackers gold pans.)

More rightly these devices are called a "miner's horn," and for many years they were not available commercially. Old time prospectors made them from a frying pan and anyone can do the same. If you want an original simply take a frying pan about six to eight inches in diameter and cut it in half. Then round off the edges and overheat it until the pan turns blue.

Commercially made horns resemble miniature gold pans and are easier to use and so inexpensive that they should be a part of every placer operators gear and indispensible for the lode prospector.

A miner's horn is used for testing a deposit for free gold values. They are operated exactly as is the gold pan, except that the panning is done with one hand, and sometimes sideways. The biggest advantage is that a sample of three or four tablespoons can be accurately tested using only the water from a canteen with enough left over to drink the rest of the day.

BATEA. The gold pan as we know it is little changed throughout the world except in South America and the East Indies. In the latter they use a pan resembling the lower half of a sphere called an "Asiatic ladle." This is of little concern to Americans as they rarely encounter it. The *batea* and some of its variations are occasionally seen and they do have some unusual advantages.

Starting in Mexico and on south, miners prefer the *batea*. Our gold pan is the frustrum of a cone, flattened off several inches below the lip. The *batea* is a true cone which comes to a point at its lower end. Its development is lost in ancient history but apparently it evolved from wooden shapes. South American primitive Indians had few skills in metal work except those of gold and silver. These metals were not suitable for tools and wood and stone were the principal raw materials which the early miners used to construct their tools. Probably the *batea* was first made from pieces of wood which were shaped by stone.

The *batea* is extremely efficient in the hands of an expert. I have tried them but years of using our type of gold pan have made me partial to the North American style. This is not to downgrade this tool. Scientifically speaking, I honestly believe that the *batea* will recover more gold than the pan if it is correctly used.

Bateas are used somewhat different than the gold pan. Cleaning the pebbles and sticks out is done much the same, by stirring and sorting. When the sand is of equal consistency, the operation becomes quite different. There is very little swirling necessary until the sand being concentrated is down to about a cupful. Instead, the *batea* is moved up and down sharply to suspend the sand in water. As this is done the silt and lighter sand is allowed to float off with the stream current or poured out if there is not enough water.

When the volume of sand is a cup or so, the miner then starts swirling the *batea* much in the manner of conventional gold panning. The lighter sand is then separated from the black sand and gold, then pushed over the side with the hands. Only a small amount of water is necessary for these final operations and care should be taken to see that no black sand or gold goes out of the pan at the same time. When the concentration is completed, the gold and high value black sand is washed over the side into a separate container.

As in gold panning the *batea* can be used above or below the surface of the water. The South American device has two advantages which are worth mentioning. One is its high sides, averaging from three to five inches. This allows the operator to "bounce" his sand much higher without the chance of losing too much over the sides. The other is the very small center section where gold values are easily concentrated without having to run the long distance of the circumference of a gold pan.

MINERAL CONE. There is one other type of panning device which is not for gold but which should be mentioned here. This is the mineral cone which is just now starting to catch on with rockhounds in this country. It is a development of the *batea* or perhaps the *batea* developed from the mineral cone. It is used in several gem localities in South America and Southeast Asia as well as other places in the world.

It is nothing but a *batea* made of woven reeds, fine screen, cloth, canvas, or some other similar material. Like the *batea* it is conical in shape and comes to an acute sharp point. If unstable materials such as cloth or screen are used, it is supported by ribs or a heavy wide mesh screen. At present, I know of no manufacturer who is making these devices commercially, but they are not too hard to construct. They vary in diamater from 15 to 24 inches. The height is about one-third of the diameter, or about six inches in depth for an 18 inch mineral cone.

The primary purpose of the cone is to concentrate larger stones which are slightly heavier or even the same specific gravity as the carrying sand. It has been used successfully to recover diamonds, rubies, sapphires and other precious and semi-precious gemstones. Since it is porous enough to let some of the finer material out, it is not satisfactory for the recovery of gold or other similar metals.

Using the mineral cone is similar to using a *batea* and the operator simply makes several sharp, downward strokes occasionally tilting the cone from side to side. This suspends the material in the cone and allows the heavier materials to go to the bottom. Even if the stones being sought are the same weight as the sand, they will generally tend to go to the bottom which seems to be against physics but it apparently works well enough in practice to permit recovery of gems. This concentrating action with a mineral cone, like gold panning, is best done under water, but can be done dry if the carrying material is very, very dry.

Unlike panning, there is no effort to wash much of the material over the side. That which floats away is permitted to do so, but the operator merely suspends his material several times, the duration of this operation depending upon several trial efforts as different sands and soils will take different times to concentrate.

Once the operator is satisfied that the material he wants is sufficiently concentrated in the lower portion of the cone, he merely turns the cone over and dumps it on the ground carefully. Then the top portion of the cone of sand (which was the bottom, concentrated part when it was in the cone) is sorted by hand for gems.

Mineral cones seem to work satisfactorily for gemstones and several rockhounds have them in their equipment. They are mentioned here so that amateur prospectors who are traveling into an area that has garnets, gem agates, sapphires, etc., can try their luck at this type of panning. If a person is proficient in panning gold, he will have no trouble in operating a mineral cone.

SLUICE BOXES. Men probably first learned to recover gold in some type of a pan. Virtually anything will work from a crock to a frying pan, but this type of recovery has one serious drawback for anyone who makes a discovery worth mining. Even the best gold panner can work only from one-third to one-half a cubic yard of gravel in one day. Supposing this were rich enough to produce a quarter ounce to the yard, which would be extremely rich in values, incidentally. With such a location it would be just a short calculation to figure that with gold at about $500 per ounce a person could earn about $30 to $55 per day. Discounting this 5% for refining and another 20% for .800 fine and the daily wage becomes about $22 to $41 per day—hardly enough gross to run even the smallest of businesses. Now assume that one individual could work from five to six yards a day. The gross goes up to about $460 to $560 per day. The richness of this hypothetical ore is of bonanza proportions but it is given here to dramatically demonstrate the economics of moving up to a bigger type of operation. To work this amount of gravel each day without a portable dredge, the amateur prospector will have to buy or build some sort of sluice.

Briefly stated, a sluice is a trough through which water passes over obstructions called riffles. Sand is shoveled in one end and the force of the current carries it over the riffles where it is agitated and concentrated. It is not intended to go further into the operation of the sluice in this paragraph, this will become obvious in the discussion of the Long Tom, rocker, etc., for the sluice is rightfully only a part of the whole operation although it is frequently used by itself.

The dimensions of the sluice are impossible to give for they are generally constructed to fit a given situation. They vary from eight to 36 inches in width and in length from four feet to sections which could add up to several hundred feet. The spacing of the riffles, their type and the height of the sides also depend upon the conditions under which the sluice is to be used.

The secret of successful sluicing is a considerable amount of water flowing through the sluice at all times at a fairly good rate of speed. Unlike the rocker, or the Long Tom coarse materials are not removed mechanically through a screen. The gravel to be worked is simply shoveled in and the water does its work. Occasionally the operator walks down the sluice and throws out the larger stones by hand. In the early days of California mining, the sluice was quite often at the bottom of a hill which was being hydraulicked and it could have been several hundred yards in length. The quantity of gold recovered from these very long sluices was tremendous. But so was the amount of sand and gravel washed through them.

When using a simple sluice box the amateur prospector most frequently simply sets it up in a stream. Water is provided by the natural current, or, if the stream is low, by building a small dam. Then he simply shovels gravel into it. As long as the sand bar or rock he is digging under is close to the head of the sluice box, they are surprisingly efficient and a tremendous amount of gravel can be worked in a short time.

Only experience can teach the amateur prospector how to use a sluice. The operation requires considerable water and there is no way to get an accurate rule of thumb as to the amount of water required to placer a given cubic content of sand in a sluice. This would depend partly upon how coarse the sand is, whether it is cemented, how fine the gold is, etc. These factors also determine the slope, or angle, at which the sluice is set, limiting or increasing the speed of the water. Two examples will show the wide variance of slope used by prospectors. In a stream a sluice will have an angle of only a few degrees more than the slope of

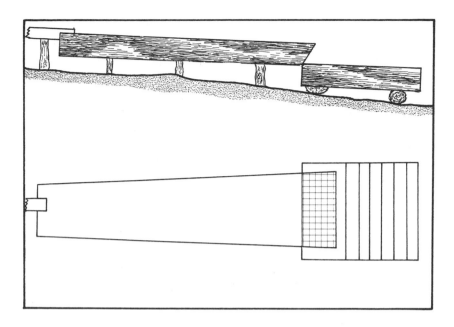

THE LONG TOM

Water turbulence using Hungarian Riffles. Diagram above riffle shows the various speeds of the water through the sluice, with the longest arrows representing the fastest speed. The circular arrows show the eddies created by the riffle allowing the sand and gold to concentrate.

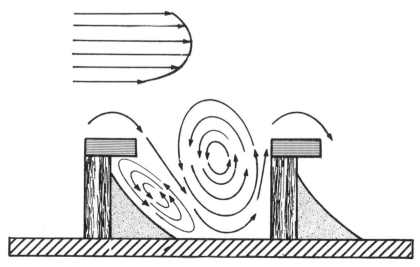

the stream. A dry washer, on the other hand, may have its sluice set up as steep as 45 degrees. Slope will be discussed further in the descriptions of the various types of devices in which sluices are used.

Despite all these difficulties of precise directions, the neophyte prospector can rest assured that his sluice will recover most of the gold if he follows a few rules used by the old time professional miners.

The width of a sluice is determined by two things: the building material at hand and the volume of gravel to be processed. Excluding the specialty sluices of portable dredges and dry washers, the narrowest practical width is about 10 inches—the widest, about 60 inches. The height of the side wall depends also on the quantity of water but is usually never less than four inches and seldom over eight inches. Riffles are so important that they are discussed separately. Smaller sluices than these are built by weekenders but they are generally used for clean-up.

The slope determines the speed of the water running through the sluice (but not the depth, which can be controlled with a head gate, or by the depth the sluice is submerged in the water). Speed is also varied with the size and type of gravel being processed. The general idea is to reach an optimum volume and speed of water that will move the larger rock in the gravel being worked. Larger boulders will accumulate near the head but these can be removed by hand when they become a nuisance—also they help to break up the gravel or clay.

Sand can be transported by water moving about one foot per second, coarser gravel requires about 1½ fps, while a one inch stone requires a water velocity of about four fps. In the field it is very difficult to gauge the velocity of water in the sluice precisely, but the observation of the size of the stone being moved by the water in the sluice will be quite accurate for practical purposes.

The quantity of water to run through a sluice must be learned by experience with the particular kind of gravel being worked. Generally, an experienced miner will start by running a depth of water which is one-fourth to one-fifth as deep as the sluice is wide. In the case of an eighteen inch wide sluice the depth would be from 3½ to 5½ inches. The depth is measured from the top of the riffles.

After several shovels full of gravel have been processed into the sluice he will then check all the riffles to see what is happening. If the sand is packing down-stream of the riffles, the velocity of the water is increased. If, in the opinion of the operator, too much sand is washing over the riffles too fast with a resultant loss in gold, the slope of the sluice or the volume of water is decreased.

The starting slope of a sluice being used in a stream is usually very little more than that of the stream. One being used on a hill or level ground generally slopes about two inches per 12 feet. Only if water is scarce is the slope increased to as much as one foot in 12 feet; in such a steep angle there is always some loss of gold over the riffles. This is usually true of any increase in slope. The idea is to increase the water velocity just enough to allow the sand to escape over the riffles while the gold will be retained. Only experience by trial and error will accomplish this.

Sluices used by professional miners were generally 12 feet long. Those used by the weekend prospector usually about six feet and often hinged so they will fit in the trunk of an automobile. A length of six feet is short for sluicing but will recover most of the gold. Since it is an axiom of placer miners that the longer the sluice, the more gold is recovered; many amateurs use a technique highly favored by hydraulic placer miners—that of connecting several sluices together with drops.

The sluices are generally stretched out in a straight line with each sluice four to six inches below the preceding one. This allows the water and sand to "drop" into the lower sluice, hence the name. This miniature waterfall is particularly useful in sluicing in that it breaks up lightly cemented materials and greatly agitates the gravel allowing the heavier materials to become separated. When using the drop, the prospector will usually discover that nearly one-fourth of the gold he recovers will be at the riffle nearest the drop.

Most sluice builders line the bottom of their sluice with a material similar to burlap or coarse canvas. In recent years the trend has been to a low nap rug. This material has an affinity for trapping very fine values in gold that cannot be captured by any other way. The material can be removed and washed in a barrel; or, burned and the ashes panned. Be careful of the more modern rugs. They are backed with a plastic material that is difficult to burn although most of the gold should wash out.

Another trick of the early day placer miners was to line the last four or five riffles with mercury which is a positive way to trap values so fine they even escape the magnifying glass. But due to the expense of mercury, the practice is seldom done today. This affinity of mercury for gold will be discussed further in the chapter on final processing.

Sluices are the major tool the professional placer miner uses to obtain gold. They can be of almost any size, length and highly refined over the basic sluices described in this section. In larger operations a dragline or a dredge supplies the gravel in tremendous quantities. However, the weekend prospector will often locate a placer deposit that he wants to work out with some form of sluice. If ample water is present he will have no difficulty in building his sluice. When the water supply is distant, or not of large supply, the problems begin. Fortunately, miners have solved these questions many years ago and the modifications of the sluice permit the recovery of placer gold even if there is no water present.

LONG TOM. Don't let the name fool you, the Long Tom is actually a short sluice. This device probably accounted for most of the placer recoveries in the very early days of the California gold rush and is still favored by placer miners working a small claim. It can be used where there is not enough water to operate an ordinary sluice, but takes more than required to operate the rocker.

The Long Tom consists of two sections—both of which are sluices. They are connected together with a drop. The first, or upper sluice, is normally about twice the length of the second, but there is no set rule about this.

At first it might appear that this is just another example of boxes put together, but there is a big difference. In the long sluice all the material is run through the gold catching sluices; in the Long Tom the first sluice serves as a place to break up the material so that it can be easily concentrated in the second, or lower sluice. The riffles in each are designed to do just these jobs and nothing more. Therefore, the riffles in the upper sluice often run lengthwise of the sluice while the lower sluice has conventional riffles set at right angles.

At the lower end of the upper sluice is a sorting device, usually a grizzly or simply a steel plate with holes drilled in it. Screens are not advisable as the abrasive action of the sand will quickly wear them out. After the gravel is broken up this keeps the larger pebbles and rocks from entering the gold recovery part of the Long Tom.

One unusal feature of the Long Tom is the slope of the upper sluice. Most often this is one inch per foot or one foot in 12. The slope of the lower sluice is adjusted along normal sluice angles. This allows the water to travel faster in the first operation and greatly aids in breaking up the gravel.

The Long Tom can be operated by one person but two men can work as much as six yards a day in loosely cemented material. One man shovels the gravel or clay into the upper sluice while the other uses a rake or fork to break it up and keep the grizzly free of larger material.

THE ROCKER. Bridging the gap between the Long Tom and the dry washer is the rocker. It is used in areas where water is available in only very limited quantities. It is little used by the amateur fraternity and this is strange. The rocker uses the least water and is probably the most efficient hand machine for recovering gold. If two men are working, the rocker can process as much as five yards a day using no more than 500 gallons of water. If the water can be accumulated in a puddle or a barrel, probably no more than approximately a hundred gallons will be sufficient for a day's operation.

Basically a rocker consists of three separate parts: screen, apron and sluice. The screen is uppermost and sorts out the gravel being worked. From here it passes to the apron where much of the coarser gold is recovered. Then the sands pass into the sluice where the rocking motion recovers the finer values of gold.

The illustration shown in this book is only a generalization of how a rocker should be built. In practice most of the specifications can be changed to fit lumber at hand or so that the rocker can be carried in the trunk of a car. The lumber should be clear and all open spaces filled with plastic wood or some other waterproof material. Unlike the temporary sluice, or the Long Tom, which can be built on the spot, the rocker is likely to get a lot of use and moving. It should be put together with plenty of screws and glue. What most people argue about is the material to use for the apron. I prefer a canvas which will hold back most of the coarse gold but still lets a small amount of water pass through.

To use the rocker, select a spot close to a source of water. If the placer ground happens to be several hundred feet from a source, water can be transported and stored in a barrel. The drawings show two spikes which fit into the flat boards to keep the rocker steady. Many other devices can be used and in some cases no boards are necessary. Not only does this keep the rocker from "walking" around, the boards can be used to establish the slope of the rocker. If the gold is fairly coarse the slope can be quite acute—as much as four inches for the length shown in the drawing. Naturally, the finer the gold, the less the slope and most rockers are worked with less than two inches slope.

After the slope has been adjusted the screen is filled with gravel and water poured over it. (Some persons use a dipper, but a pail is handier and holds more water.) Then the gravel is worked with a small rake or by hand until all the smaller material has been pushed through the screen.

Now the rocker is rocked vigorously back and forth with several quick, abrupt stops to further agitate the gravel. More water is added during this operation and the gravel in the screen further agitated. The process is carried on until no more material can be forced through the screen.

The screen is then removed and the larger stones dumped out. It is refilled and the whole operation repeated. Each time the screen is removed the pocket in the apron should be checked. Almost all of the coarse gold will be trapped here and when it can hold no more the apron should be emptied into a container for final concentration. Since there are usually fewer riffles, these too, should be checked frequently for cleaning up.

The critical part of learning to operate a rocker is the amount of water to use. Too much and the fine gold concentrates simply wash away with the sand. Not enough water will make a muddy (especially in clay) water which will not let the fine gold settle and it will also be washed away. One peculiarity of the rocker is the tendency of the black sand to harden up and make a firm surface in front of the riffle where the rest of the material simply washes over the top with no concentration. When this happens try being more abrupt when stopping the rocking motion to agitate the sand more. Sometimes nothing will work and the riffles must be cleaned more often.

DRY WASHER. When there is absolutely no water, the only way to recover gold is with a dry washer. Unless you live in the desert areas of Southern California, Arizona or Nevada, the chances are that dry placering is completely foreign to you. However, people who are weekend prospectors and live in dry areas have recovered considerable amounts of very pure gold from some of the more productive dry placer gold districts. Dry placering can be described very simply by telling how the early Mexican miners who discovered most of the dry placer deposits got their gold. On a hot day with a little breeze blowing, two men would shovel dirt into a blanket, pull it tight between themselves and bounce the dirt and dust up and down. The breeze carried off the dust and smaller particles while the remaining sand was gradually concentrated in the blanket until only small quantities were left. These were removed, taken to an area where there was water and the fine gold values were then recovered by gold panning.

The modern weekend prospector doesn't have to depend on a partner to help him shake a blanket. There are several machines on the market and one can be built in the backyard without too much difficulty. There are also plans available from several sources for building a dry washer.

Most of the commercial versions are adaptations of the original Mexican dry washer used by old time desert miners. These operated off a crank and bellows but gasoline engines have taken much of the labor out of dry placering. Air is supplied from a compressor or large fan.

Generally a manufactured dry washer consists of three major parts. When assembled for traveling it resembles a box. When set up for operation the parts are separated several inches by supporting arms. The top box is a coarse screen into which the material to be worked is shoveled. Through this screen it falls onto another screen where it is further sifted into much smaller particles. From this second screen, it falls onto an inclined trough with three, four or five riffles about a half to a full inch in height. The spacing of the riffles in the dry washer is much further apart than the sluice box for much more efficient operation. The lower sluice is always covered with the gunny sack or canvas onto which the riffles are nailed. Below the sluice is a bellows or a compressor which forces air through the canvas or gunny sack raising the fine particles which have been sifted through the first two screens. The volume of air is not considerable, just enough to lift the dust and lighter particles up over the riffles and let them work down towards the end where they tumble off as waste. It is easy to see that if too much air is used, all the material, including the gold, would eventually be pushed up and over the ends. If a compressor is used it should have a valve that will permit complete reductions in the amount of air used.

The screens and riffles are hooked together usually with a Vee belt attached to pulleys attached as eccentrics so that when the lower pulley is turned all three vibrate.

This was the common type of dry washer that was universally used until recently. Then, Keene Engineering developed what they call an electrostatic concentrator, that may greatly simplify dry placering.

Keene's dry washer is extremely light, about 54 pounds, and highly portable. Made of aluminum and plastic it is quite durable. It gets its name from the concentrator unit which is made of high impact plastic. There are more riffles than commonly found in a dry washer. Beneath these is a static charged cloth which aids in holding the gold particles while the air blows away the lighter material.

One of the major advantages of Keene's new dry washer is that it provides warm air to help dry the sand if any moisture is present. The blower cage is so

148

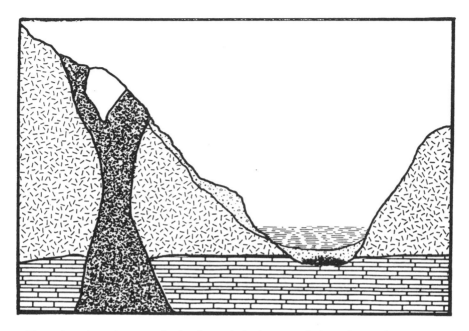

Top drawing shows a lode deposit being eroded and creating
a placer deposit. Bottom reveals a Tertiary river being eroded
to create a deposit. Both drawings are greatly exaggerated for
clarity. In nature, the deposits would be much further away.

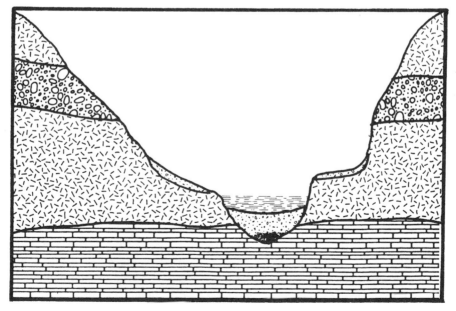

designed that heat from the engine is transferred to it warming the air which is produced for the concentrating table.

I have met many people who have been extremely disappointed by dry washing on weekends. My own experience in this department is very small and is gained mostly from acquaintances who live in desert areas. Most of them have large collections of gold they have recovered and most have little difficulty in getting gold everytime they go out. Their particular secret, and one that most neophyte dry washers overlook, is to pick the right time of the year. It should be the hottest, driest time of the desert summer. What most people seem to forget is that after you dig down a few feet there is some moisture in the earth—and any moisture at all makes dry washing difficult if not impossible. Like all other types of prospecting it is necessary to dig deep to get the values from a dry placer. One of the more successful tricks in dry placering is to bring the richer material to the surface and spread it out to dry for a day or so. When it is run through the dry washer it works much better.

The time of the year to dry placer on the desert is not the safest time to be in this country. *Be absolutely certain you know what you are doing!*

DIP BOX. In locations where there is not enough water for sluicing, and the ground is too level to permit a good slope for a sluice, a dip box can be used if a rocker is not available. Briefly, a dip box is simply a short sluice mounted on legs so that it is about waist high. A screen is mounted over the head of the sluice to sort the gravel. The slope is determined by the height of the front and rear legs.

A dipper or pail is used to dip both water and gravel into the dip box where it runs down the slope, the gold being concentrated in the sluice. This is a rather crude way of getting gold but often effective in remote areas. It is usually a makeshift, done on the spot affair, when no other type of sluicing is possible.

PUDDLING BOX. Gold is not always found in sand and gravel, it is often located in clay. Recovering gold from clay in a sluice is often a difficult proposition. The clay has a tendency to "ball up" and literally roll down the sluice. These balls can actually pick up gold that has already been concentrated and carry this gold out the waste end of the sluice.

Therefore, it is very important in this type of operation that the clay be broken up. This can be done in a puddling box which is simply another sluice attached to the head of the other sluices in the line. In operation it is similar to the Long Tom, and the material is simply broken up with a hoe, rake, fork or similar tool.

THE UNDERCURRENT. The undercurrent is a product of diminishing placer values. It was invented in pioneer days to recover every last color, and undercurrents have been known to recover values as fine as 1/2,000 of a cent (and this at $35 per ounce). The undercurrent is a complex piece of equipment, but it can be built portable and transported to a location. It would be used in a major operation where the weekend prospector may discover a sand bar that is going to pay out all summer. Otherwise, it simply isn't worth building at all.

To be successful an undercurrent requires a lot of sand and raw material; plus a lot of water and volume because it runs from one to seven sluices at the same time.

At the highest altitude is the main sluice, ordinarily put up with common riffles. At the point where the water enters and the material is shoveled in, is a form of pole riffles, rock riffles, log riffles, or railroad track to break up the material. From here it passes to the entrance of the main sluice. Just before the material being sluiced enters the main sluice it passes over a grizzly. Usually this grizzly has a grid pattern of one-quarter inch and it is often as small as an eighth of an inch. Naturally, a lot of the material will pass over this grizzly into the main sluice. Here the coarse gold and some fine values will be recovered.

Through the grizzly, water and finer parts of the sand pass into a feed box. The feed box then transports the sand and water from three to six additional sluice boxes which are set up with riffles. These are called the undercurrent and since it is lower than the main sluice box and a smaller current of water runs through, it is aptly named.

It takes a tremendous quantity of sand going into the main sluice to get much sand down into the undercurrent but that sand which is concentrated here receives a much gentler sluicing action and extremely fine particles of gold can be recovered. Each sluice is covered with a blanket, gunny sack, canvas or other similar material and the riffles are ordinarily constructed lower.

Amateur prospectors will rarely, if ever, need to build an undercurrent. However, in a large operation, or when many fine values are being lost, it is the only answer.

GRIZZLY. To a miner a grizzly is simply a screen which sorts out larger pieces of rocks so that only the finer sand and gravel reaches the sluice. A grizzly can be simple iron straps placed from a quarter to a half inch apart. One of the most popular, and longest wearing, is simply holes drilled in a piece of eighth-inch steel. Commercial screens of one-quarter mesh are frequently used, but these have a habit of wearing out after a short period.

The grizzly must be cleaned out frequently since large stones will build a dam holding back the sand and gravel from entering the sluice. Almost any prospector has a special feeling for cleaning out the grizzly and does the job with utmost care. It is here those rare nuggets are occasionally found and who wants to throw one of these prized finds away.

RIFFLES. Riffles are not really a tool, but they are such an important segment of sluicing that they deserve considerable discussion by themselves. Among miners there is a lot of disagreement as to which is the best riffle. The answer, of course, is that there is not a "best riffle." Each riffle works a little different and should be selected for the nature and composition of the material being sluiced.

Riffles can be as simple as nailing a common lath to the bottom of a board, or as involved as welding angle iron to steel plates. Each type of riffle has its advantage and sometimes more than one type is used in the same sluice.

Riffles are constructed of almost any type of material imaginable with the most common being wood. It should be remembered that with long use, wood is quickly ground up by the tremendous quantities of sand and gravel that pass over it. For a sluice to last a long time, steel, aluminum or any other metal will work well and stand up for many years.

The main purpose of the riffle is to create an obstacle for the sand or gravel being worked. At the point of this obstacle, the water forms an eddy, (or as the old time miner used to say, it "boils") allowing the gold to settle near the riffle. All obstacles to a stream of water will do this to some extent. Some shapes are far more efficient in creating eddies and as the fineness of the gold increases, more complex riffles are used to allow it to settle.

There is no way it can be predetermined how many riffles will be needed or how far apart they should be. This is determined by the composition of the gravel and the richness of the gravel and the fineness of the gold. This poses something of a problem for the amateur prospector since his sluice is usually built at home and transported to the field. Without knowing the nature of the gravel, nor the type of placer he will be working, he has to guess far in advance how his riffles should be placed.

Most of the sluices I have seen were from eight to 12 inches wide with sides from four to six inches high. The riffles were from eight to 12 inches apart. Most weekenders will have to transport their sluice in the trunk, or on top of a car, so they are relatively short, usually about four to six feet. Two sluices can be built and used in conjunction to give added length.

It is probably a good idea to try the sluice out somewhere near home and observe how well it works on sands before you take it to the field. The riffles should be watched carefully to make certain they do not pack hard with black sand or clog. If this happens the gold will wash right over the riffle. If the eddies are too violent much of the gold will wash over the top along with the lighter materials.

Common Riffle. This is nothing more than a slab nailed cross ways on the sluice box. It can be as small as a lath or as big as a two-by-four. More often, they are a piece of one-by-two common lumber.

Zig-zag Riffle. This is a variant of the common riffle and is used generally to break up material before it enters the recovery sluice. It is not a very efficient gold catcher. As can be seen from the illustration, the current is swept from side to side much as a stream meanders. Usually the sluice with the zig-zag riffle has a greater slope letting the water travel faster. This tends to swish the gravel back and forth sideways and is quite effective in breaking up lightly cemented material without too much hand agitation.

Hungarian Riffle. Hungarian riffles are made by screwing a piece of iron bar stock about one inch wider than the riffle to the top of the riffle. The overhang is usually on the down side of the riffle although it can be on either side This

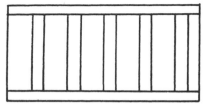

COMMON RIFFLES

ZIG ZAG RIFFLES

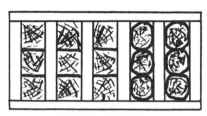

BLOCK RIFFLES

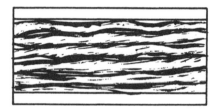

POLE RIFFLES

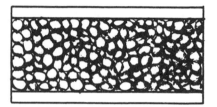

ROCK RIFFLES

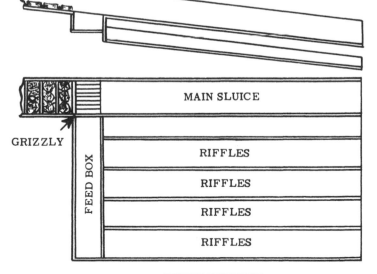

MAIN SLUICE

GRIZZLY

FEED BOX

RIFFLES

RIFFLES

RIFFLES

RIFFLES

UNDERCURRENT

configuration creates quite a bit of swirling action and helps settle fine gold. This type of riffle was used extensively in the dredge (bucket) operations in the peak mining periods and many claims were made for its ability to recover fine values. Another method of making the Hungarian riffle is simply to use angle iron.

The Hungarian riffle has advantages and disadvantages. To work right the riffles must be set very close together, usually about two inches apart. Cleanups are difficult and there is a tendency for them to clog or fill up if much black sand is present.

Angle Riffles. Relatively a late comer, this type of riffle is the favorite of the portable dredge prospector and is remarkably efficient in recovering very fine gold. For dredges they are set close together, only a few inches apart usually. The large volume of water produced by the suction pump keeps the water boiling continuously over these riffles.

Since there is very little classifying of the material scooped up by the suction tube, the closeness of the riffles helps to keep bigger stones out of the riffles. Also, as these stones bounce over the riffles, they jar the material being worked to help keep the sands being concentrated in suspension.

The problem of cleanup is eliminated since the sluice of the underwater dredge is fairly small and can be simply turned over to dump the concentrated material out. Some come apart and can be cleaned super efficiently.

Pole Riffles. Weekend prospectors will seldom use the transverse type of riffle unless it is to break up gravel. One of the cheapest and most efficient is the pole or log type of riffle. These can run either way of the sluice but early miners preferred the lengthwise placement. Many were the arguments early placer miners had about pole riffles, but those who used them claimed that the circular shape caused the water to have its best "boil" and settle finer gold particles.

Rock Riffles. Rock riffles were developed for the same reason as were pole riffles—they were at hand, simple and cost nothing. Probably they were the first riffle used by primitive man to obtain gold. It is difficult to state with accuracy what size or shape of rock should be used since rock sluices are usually constructed on the site from what is at hand. Placement of the rocks also varies with personal preference. Some miners used big rocks widely spaced and ran a considerable volume of water past them to break up cemented material. Others actually quarried them in the shape of riffles.

Rock riffles are not one of the most efficient methods of obtaining placer gold but they do work. One big advantage is they can be built on the spot and often nothing but the rocks are required.

Other Riffles. There is no end to the types of riffles and the materials used. For example, if they are available, railroad tracks make excellent riffles either lengthwise or crosswise. Railroad ties have been used, as have cut up tires, old car parts, or nearly anything you can name to create an obstacle. The weekend prospector who discovers a rich bar and has only his pan along, should start looking for something to construct a sluice with. If he uses his head he won't have to take the chance of someone else finding his bonanza before he can get back with his sluice.

SLUICE COVERINGS. To be most efficient some section of the sluice should be lined on the bottom with a material like canvas, burlap, rug or corduroy. With this material it is possible to recover the finest particles of gold. To be efficient, the lining material should be easily removed to clean up—a slick trick is to simply staple it down with a heavy duty stapler like those used to install wall board.

There is much difference of opinion on how much of the sluice to line and what material to use. Experience, and the fineness of the gold will be the only way for each individual to find out. Generally, the last couple of riffles are lined, although some sluicers line the entire box.

At intervals, and always when shutting down where the covering material will dry out, it should be removed and cleaned. Swirl it around in a tub of water and pan the material which settles to the bottom. When the lining material is no longer usable, burn it and pan the ashes.

During the sluicing operation watch the lining to see that it does not plug up or get slick with black sand or clay. If this happens, the fine gold will simply wash over it and it is time to switch to another material.

OTHER TOOLS. There are many other tools which can make gold recovery easier and more efficient. Purchasing or building some of them will be expensive and the weekend prospector will have to be very successful, and probably become a weekend miner, to justify their acquisition. Others are simple and can be built for little cost. Whether the active amateur wants, or even needs, some of the tools described from here on is pretty much up to him. They are described so that each can make his own choice.

RETORT AND CONDENSOR. A retort is a device in which mercury is heated to the vaporizing point leaving a residue of gold for further refining. The condenser then brings the mercury back to its original state. An explanation of amalgamation, the process of evaporating mercury is complex and potentially dangerous. It is the the subject of a lengthy discussion later in the final concentration methods chapter of this book.

ROTARY MILL. A rotary mill is used to reduce hardrock ore to a fine mesh so it can be panned or sluiced. Commercial mining operations use these in tremendous sizes, but there are smaller, portable sizes. Weekend prospectors who find a good, small lode deposit they want to work will find many portable mills advertised in mining journals. Ordinarily these weigh about 160 to 200 pounds and can be operated with a hand crank. They crush ore to about 200 mesh, fine enough for the pan or sluice.

MORTAR. A cast iron mortar and pestle are handy for the lode prospector. They come in several sizes and are used to crush ore so it can be panned. For prospecting, the six to ten pound sizes are best. Larger ones weighing up to 100 pounds can be used to work small deposits by hand.

In operation, quartz, or other country rock, is placed in the mortar where it is crushed by the cast iron pestle. It is hard work but even by hand, the ore can easily be crushed to 100 or 200 mesh, making panning an easy operation.

SORTING SCREEN. Weekend prospectors frequently use a sorting screen to classify material before panning or sluicing. They are easily constructed from wood frames to which a quarter or half-inch screen is nailed. Mostly they are used in panning or very small sluicing operations. If much material is to be classified, it is better to build a grizzly out of heavier material, the screen won't hold up under heavy use especially if larger rocks are thrown in.

SIEVES. Geochemical sieves come in various meshes and are reasonably priced. Useful to the commercial assayer, the weekend prospector would probably want them for sorting his gold into different fineness. Some amateurs like to display their collection as to the different degrees of fineness.

PICK. An absolute must for either the placer or lode prospector, or, the dredge operator. The minimum requirement is the small, portable kind which breaks down and can be carried on a belt. These are usually found in surplus stores and are excellent for long walking prospecting trips. Every serious amateur should have a large miners pick. Not only are these useful for breaking up loosely cemented material, they can be used to break many kinds of rock. There will be many times that a shovel will not be enough. A good example is trying to dig in dried adobe.

SHOVEL. A miner without a shovel is like a farmer without a plow. Outside the pan, it is really the one basic tool for every type of prospecting or mining, whether it is placer or lode. As the weekend prospector becomes more proficient he will probably end up with a collection. For prospecting I have the GI issue with a sheath which balances the GI pick on my belt. With a canteen full of water it makes quite a load for walking but when you find a likely spot, you have everything you need to get deep where the gold will be.

Naturally, I have a regular miners shovel. My gear includes a long and short handled one and frequently I take them both. Sometimes the short handled one is quite handy in situations where the longer handle is unwieldly.

TWEEZERS. There are some things which I consider almost an absolute necessity. For panning, one of these is a good pair of tweezers. Not the type ladies use for their eyebrows, although these will work, but the longer, commercial kind. These are made for many industries, but my favorite are those made for jewelry workers. They are about six inches long and can be found in any rock shop that also carries hobby equipment. I usually file the ends down to about one-sixteenth inch wide and the insides perfectly flat.

Frequently in panning, snifting, crevicing or just plain classifying sand, a small fleck of gold will appear. It is easy to remove this with the tweezers and pop it into a small vial or bottle. Both of these are always in my shirt pocket when I am panning or prospecting.

MAGNIFIERS. The easiest way to identify gold is by looking. In panning, especially when prospecting when the sand comes from relatively near the surface, the colors are likely to be few and very small. It is sometimes possible to see gold grains with the naked eye, but a magnifying glass makes it far easier.

There are all types of glasses and any good magnifier will work. The main thing to remember is that the greater the magnification, the closer the lens must be to the object to be in focus. A ten power, for instance, can only be about a half inch from the object to be clearly seen, making this power difficult to use in the pan. Four to six power is ample for field work and can be focused far enough back to be convenient.

I have several glasses and take more than one power with me on a field trip. Sometimes in prospecting it is important to know whether the grain of gold is smooth or has rough edges. To determine this, a ten to 20 power glass is helpful. If the grains are very small, I use a fifty power hand microscope.

One other handy magnifying glass is the comparator. This is a glass with its own stand. At the bottom is a piece of plain glass with measurements etched on the surface. Although they are only about six power, comparators can be used to measure distances smaller than a sixty-fourth of an inch. This gives the prospector an accurate indication of how fine the gold is and helps him determine whether an undercurrent is necessary, or whether some other type of riffle might be more efficient.

Don't worry about buying a whole box of different magnifiers when you begin. Outside of the four to six power glass used to pick out small colors in the pan, the others are mainly a luxury. They can be added later as the prospector acquires more experience and equipment.

MAGNETS. For separating magnetic or iron particles (black sand) from gold, a good magnet speeds the process up so much that it might be called a necessity. There are many excellent ones on the market and the size is determined by the use. Those for the waxed paper method need not be as strong as those used for the plastic gold pan. Don't buy a cheap one, as they do not last long enough. Most magnets have a bar of steel which fits between the poles. Keep this on the magnet when not in use and the life will be greatly increased. Also remember that sudden shocks will shorten its effective life, so don't drop the magnet and pack it so that it will not bounce around when it is being transported.

SNIFTERS. Snifters are used to suck small concentrations of gold or black sand from hard to get at places between rocks. Some suppliers have such devices but a good substitute can be found in the syringe department of the local drug store. Ear syringes make a good small one and larger syringes are available. One of mine was pirated from the kitchen where it had been used to baste a turkey. Snifters are used both under water and on dry land.

CREVICERS. A crevicing tool is used to reach down into a narrow crack in rocks and pick up a small amount of gold or black sand that has concentrated there. There are some commercial models but weekend prospectors need only to let their imagination run wild. My first was a spoon bent at right angles and attached to a wood dowel. Sometimes this would be too short, and sometimes too long. The dowel had to be half an inch in diameter to have enough strength and even this was too thick at times.

Finally the dowel broke and I got smart. I purchased two rifle cleaning kits. By combining them I could reach nearly four feet. I worked over several spoons and drilled them so they could be attached at the bottom with a small screw. The whole thing can be packed in a small package and gives me about twice the depth. Don't forget to reshape the spoons so they are deeper and narrower.

COMPASS. For some prospecting trips a good compass is a necessity. This is especially true if the trip is into the back country or if the location must be found from a topographical map. My experience is that very cheap compasses are useless and inexpensive ones not worth the trouble that they cause. Even a well engineered compass will sometimes not work in certain mineralized zones. A good compass should have a dampener to minimize vibrations and should give readings to about ten minutes. Always remember that it is one of the most delicate instruments a prospector carries and care should be taken to keep it from being abused.

ALTIMETER. This is sort of a luxury but will help in trying to find a stream from a topographical map. There are several, very accurate pocket altimeters but these are also very expensive. For most prospecting trips a modestly priced altimeter will do the job.

CLINOMETER. A clinometer is a device for measuring angles. The weekend prospector will seldom have use for them since most of his work is in placering and prospecting. In mining they are used to determine the dip of a vein. Many compasses have this feature and a relatively accurate device used by carpenters to measure angles to 45 degrees can be purchased in most hardware stores for a few dollars. This would probably be sufficient for the average person as he would only be making measurements on hills or an old mine to satisfy his own curiosity and extreme accuracy would not be needed.

DRILLING HAMMER. Prospecting for lode always means some rock drilling. A good portable hammer for this is about the size of a maul. Rockhounds call them a rock cracking hammer. Each individual should select a weight that feels right for him. After about an hour of drilling, the weekend prospector will know why these hammers are offered in many weights. Too light and the drilling goes on forever, too heavy and a man's shoulder gives out in a short time.

ROCK HAMMER. The most useful hammer for the weekend prospector is the rock hammer used by rockhounds. It can be purchased in any rock shop and used for a multitude of things. It will serve as a pick or a rock crusher. It will dig out soft rock and can be used to split shattered rock with good cleavage. *Never use it as a chisel by hammering it with a maul.*

CHISELS AND WEDGES. For the lode prospector the most useful accessories he has will be his set of chisels and wedges which seems to grow as the years pass. They are used to split rock sections to expose veins or break off pieces of rock to be ground up for sampling. Many varieties are available from supply stores or they can be made up by a blacksmith out of high carbon steel. Some have square points called gads, while others are shaped like a conventional cold chisel. Wedges can be those used for splitting wood since they are rarely used to cut, only split rocks. Chisels can range from six inches to several feet in length. Weight is a major consideration and most chisels are less than two feet long.

DRILLS. Rock, or star drills will be necessary only if blasting is going to be a part of the lode prospectors operation. While there are power drills available, these will rarely be used by the person who makes a hobby out of mining. Drills are best purchased commercially and sharpened often. Hand drilling a hole in rock is a long, tedious job and, fortunately, discourages those who might otherwise use blasting indiscriminately.

REFERENCE COLLECTION. A reference collection of minerals is a handy thing to have on a prospecting trip, especially if the professionally identified minerals are augmented with specimens collected in the area being prospected. They range in price from a few dollars to a few hundred dollars.

SCALES. Many amateur prospectors have a set of miners scales. These can be sophisticated sets which will accurately weigh a hair from your head or simple scales that are relatively inexpensive. Balancing scales are best but there are several pendulum types which work well although their capacity is limited. Since most amateur prospectors seldom sell their gold, the scales are generally used to satisfy their own curiosity as to how much gold they have.

This is no means the complete list and there are dozens of other gadgets and tools that will occur to the inventive weekend prospector. Some will be as simple as discovering that a whisk broom is a handy thing to take to brush concentrated sands from depressions in high water rocks. Others will be more complex—such as the crevicing device a friend of mine made up by crossing a vacuum cleaner with a portable generator.

17

FINAL CONCENTRATION METHODS

No matter what method the weekend prospector uses to recover gold, he will rarely have a large quantity of pure gold for his efforts. In panning, the gold can be separated quite efficiently from the black sand so that the end result may be as much as 90% gold.

In sluicing, it is another matter. Tremendous quantities of gravel are run over the riffles before the cleanup is made. Even the most efficient sluicing operation will still accumulate considerable black sand with the gold. There is rarely time to further concentrate this material and it is usually cleaned out from the sluice and stored in a glass or plastic container until it can be processed further.

Then, there is the matter of very fine gold—that which is in particles too small to be seen with the naked eye. Fine gold is extremely difficult to separate by normal methods. Even if you are panning and feel that you are making a good separation of gold from black sand, it is wise to examine the black sand under a glass to see if there are minute particles remaining. If there are, put all the black sand in a separate container and process it later.

This type of concentration is usually done after the panning or sluicing is through. Many amateurs take their black sands home to process where conditions can be more controlled. Professional miners often did it in the evening, after the day's work was done. Since it is the last thing done with the day's recovery, it is called "final concentration." It simply means to get the last miniscule part of gold and eliminating any other foreign material.

There are two steps in final concentration—a mechanical type of separation and a chemical type. Mechanical methods are used as the first process.

The first of the final concentration methods used depends a lot on what process was used originally to concentrate the river sands. Panning and dry washing concentrates are similar, very high in gold with small quantities of black sand. With this type of concentration you will most often switch immediately to the chemical or magnetic method.

On the other hand, if your concentration comes from a sluicing or dredging operation, the chances are good that there is quite a quantity of black sand compared to the gold. Or, if you have fine values in your concentrate, the quantity of black sand will also be considerable.

If the quantity of black sand concentrated is a cup or more the first step is to re-pan it. The procedures are the same as when panning in the stream. However, you should use a smaller pan. In the field an experienced prospector will use an 18 to 24 inch pan so he can work more gravel in a given time. Now you should switch to a pan no larger than 12 inches. There are a couple of reasons for this.

First of all is the fact that the concentrate is now highgrade ore. It might have run a pennyweight to the ton in the stream, now its value could run up to several ounces to the ton. The second is that you will be working with reduced quantities of sand and it is easier and more efficient to work with a smaller pan.

If you are working in the field, be sure to find a quiet spot in the stream where you will lose less of the sand to the current. The idea in this panning is to get as much of the gold as possible before going to the next step. Separating the black sand and gold is exactly the same as in the original stream panning except at this stage of the game none of the black sand will be thrown away—it will all be processed once more.

If, after you have made this final concentration, the particles of gold are now powder fine, the few remaining particles of black sand can be separated with a magnet. For this you are going to need a magnet, some waxed paper and a few rubber bands.

You will also need a container for your gold concentrate. A low, wide

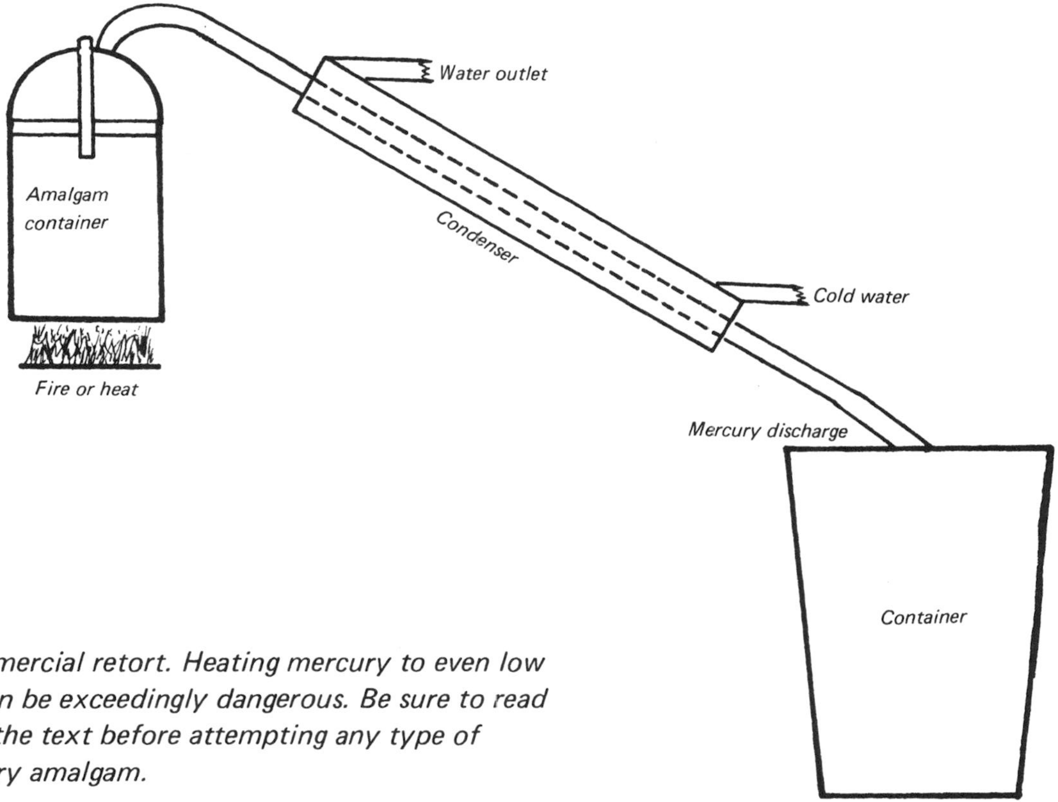

Basics of a commercial retort. Heating mercury to even low temperatures can be exceedingly dangerous. Be sure to read the cautions in the text before attempting any type of retorting mercury amalgam.

Amalgam container

Fire or heat

Condenser

Water outlet

Cold water

Mercury discharge

Container

mouthed jar like those used for creams is ideal. It is not too deep for the magnet and the smooth inside makes getting the gold back out easy.

Put the concentrate in the jar and let it dry thoroughly. To separate the black sand from the gold requires that both the gold and sand be absolutely dry. During the drying procedure tap the jar occasionally to make sure that residue of river silt does not cement your concentrate.

The magnet should be prepared by covering it completely with the waxed paper using the rubber bands to hold it together. Now move the magnet down towards the dry concentrate until the black sand particles adhere to the waxed paper. Then move it up an inch or so and tap the magnet sharply. This will dislodge any gold particles that were trapped by black sand particles and let them fall back into the jar.

The jar should now contain native gold as pure as it is possible to refine it outside an assay office. This is usually the most valuable of gold and if you have a good quantity of it, native gold can often be sold at a premium to jewelry and novelty manufacturers.

The waxed paper on the magnet should have several particles of black sand and a little silt-dust adhering to it at this time. This should be removed over the other container which has the larger quantity of black sand which is now ready to amalgamate.

Due to the high cost of mercury, amalgamation is not done very often any more. However, if there are fine values it is the only answer. Also, if you have several pounds of concentrate from a sluicing operation, it is often tedious and time consuming to re-pan it in the small pan.

There are several ways that amalgamation can be done. The most common method is to use a regular pan. About an ounce of mercury is placed in the pan along with several pounds of concentrate. The pan is then placed under water and agitated in the same manner regular panning is done. When you can see no more free gold in the concentrate, or when you are convinced that all the heaviest concentrates have been passed over the mercury, the black sands are poured off in the same manner as in regular panning. Be careful not to let any of the mercury escape over the side in this operation. If the mercury has separated during the panning, it will normally rejoin.

Copper pans are the most efficient collectors of fine gold and will get particles too fine to be seen through a six power glass. Coarse river gravel cannot be treated this way but black sands will work. The copper pan is first roughened with emery cloth and then rubbed with mercury until it has a bright silvery surface. Smaller quantities of concentrate are then panned over this and the big, wide surface gets the gold almost immediately.

When the mercury collects enough gold, it is scraped off with a putty knife or some similar tool. Then fresh mercury is added to the pan and the process begun again.

Mercury is often used in sluices where it is generally placed in the first riffle so that in the event it flows over, it will be trapped in the succeeding riffles. Cleanups are more frequent when mercury is used to make certain it does not escape the sluice.

After mercury has accumulated fine gold in any one of the proceeding manners, it is called amalgam and must be treated to form what is called a "button." The button is later treated by a refiner and the result is pure gold.

Before describing how to make a button, a few words of caution are necessary. *Mercury fumes are exceedingly poisonous and many miners have died from inhaling them.* When heated to the evaporation point (675 degrees) a heavy, white smoke will appear. *But, when heated only to the temperature of boiling water, it will volatilize and start giving off near invisible fumes.* Many persons have suffered permanent damage to themselves, and even death, because they thought the mercury was not vaporizing simply because they did not see the heavy white smoke.

Unless you are working for several months and need to re-use your mercury, retorting, or heating the mercury to make a button is not recommended. The best solution is to take the amalgam to a refinery and let them do it professionally. But, if you must, there are several ways—some of them not very safe.

The simplest of these is to just put the amalgam on a piece of sheet metal and heat it over a fire. The old timers often used a frying pan, but it goes without saying that the frying pan should not be used to cook food later. This method will produce a sponge-like button, but will result in losing all the mercury through evaporation.

Another open-fire procedure is the potato method. A medium sized potato is split in half and a small depression carved out in the middle with a spoon. The depression should be much larger than the amalgam. Then the amalgam is placed on the sheet metal or frying pan and the potato placed over it. The pan is placed on the fire and heated for about 20 minutes or longer. Fumes from the mercury will be escaping from beneath the potato so this operation is not entirely safe either.

After the potato has cooked long enough, it is removed and if all the mercury is gone from the amalgam, the job is done; if not the process of heating is carried on for several more minutes. When the button is ready it is removed and the potato can be crushed and panned to recover the mercury.

These are the two common methods of transforming small amounts of amalgam in the field. If either is ever used, it should be over a fire in the open with the wind blowing the mercury fumes away from you. *Never heat mercury inside a closed room.*

If you are working steady through an entire season, you will probably purchase a retort. These devices heat the mercury quickly, cool the vapors with water and recover almost all of the mercury so it can be re-used. They can be purchased from mining supply houses and full instructions should accompany them.

a

abrasion. Wearing away by wind, water or other type of friction.

accidental. In volcanic rock—a broken fragment not of the magma of the eruption.

accretion. A process wherein inorganic bodies increase in size by the addition of new particles to their outside periphery

acicular. A mineral which incorporates within itself fine, needle-like crystals.

acid. Sour, sharp or biting to the taste; in rocks, one in which the silica predominates; in geology, a test for the identification of rocks.

acidic. Used to describe igneous rocks which contain more than two-thirds dioxide os silicon.

acre. 43,560 square feet of land in the United States; (4,840 square yards or 160 square rods).

acreado (Mexican). Any gray or steely colored ore; a gray, copper ore.

acre-foot. The quantity of water, overburden, soil or other material required to cover one acre, one foot deep. (An acrefoot contains 43,560 cubic feet.)

acre-inch. The quantity of water, overburden, soil or other material required to cover one acre, one inch deep.

adamatine. Diamond hard; commercial name for drilling bits; having a luster like a diamond.

adit. A horizontal passage driven from the surface to work a mine or vein. (A tunnel, for example, would be a passage driven to the opposite side of a hill or mountain.)

adobe. Clay and silt mixed together. Mostly applied to deposits found in the desert basins of the southwestern United States and Mexico where the material is commonly used in the manufacture of bricks.

AEC. Atomic Energy Commission.

aeolian. Obsolete; a word seldom used which is synonymous with eolian. Refers to material deposited by the wind.

affluent. A tributary stream or river.

A-frame. In mining, two studs or poles supported in the shape of the letter "A."

agate. Silica in which variegated bands are presented in various colors.

agglomerate. Breccia composed largely or entirely of fragments of volcanic rocks.

aggregate. A collection; to bring together any kind of minerals or rock fragments in a substance faintly resembling concrete.

alkali. A Substance which neutralizes acids; calcium, potassium or sodium.

alkali flat. A relatively flat area in an arid region where salts have become concentrated by evaporation.

alkaline. Frequently applied to areas having alkali qualities; an ambigious term.

alloy. A metallic substance composed of two or more elements.

alluvial. Minerals dispersed by action of wind, water or gravity; often used erroneously to describe any overburden containing minerals.

alluvial fan. Used to describe the fanning aspect of a stream as it leaves a mountain to disperse in a valley; also used in desert areas to describe the effects of a mineral breaking off the steep sides of mountains and being spread by later effects of water or wind.

alluvium. Sediment, clay, sand, gravel, or other minerals transported and deposited in fairly recent geolgic times.

alpine glacier. A stream or sheet of ice located high in a mountain valley and fed by streams or compacted snow.

altered rock (or mineral). A rock or mineral that has undergone chemical changes during a long geological process.

altitude. Height above sea level usually measured in feet or meters.

amalgam. A physical alloy of mercury with one or more other metals.

amalgamation. Process of passing mercury over a highly refined ore to recover a valuable metal such as gold.

amber. A very hard fossilized plant resin. Yellowish in color it is often used for semi-precious gem stones.

amethyst. Transparent quartz, purple or violet in color; very popular semi-precious gemstone.

amaporo (Mexican). Continued ownership of mining claims proved by the employment of a specified number of men.

amorphous. Not crystalline; rocks or minerals having no defined crystal structure.

amygdale. Vapor or cavities in igneous rocks which have become filled with different minerals at a later geological period.

andalusite. A silicate of alumina found in schists and gneisses, often used as a semi-precious gemstone.

anhydrous. Literally, without water; applied to minerals which do not contain water of crystallization or water of chemical combination.

anthracite. Hard, very pure, lustrous coal containing a high percentage of carbon.

anticline. Strata which dips in opposite directions from a ridge or an axis.

antiqua (Mexican). An old, deserted mine or workings.

apatite. A soft mineral which occurs in hexagonal forms; popular with rock collectors.

aphanitic. Used to describe rocks in which the crystal structure is too minute to see without magnification.

aqua regia. Corrosive chemical compound made of three parts hydrochloric acid and one part nitric acid. Dissolves gold and platinum.

aqueous rocks. Refers to rocks deposited by the action of water; sedimentary rocks are of this type.

aquifer. A geological formation capable of producing original water as opposed to natural water bearing stratum in the normal water table.

Archezoic. Early pre-Cambrian times, oldest recorded geological times; as modern geological science is developing, this term will soon be obsolete.

argentiferous. Silver bearing ore; often applied to galena.

arenaceous. Rocks composed of, or derived from sand.

arete. Acute, rugged mountain range between two mountain ranges.

argillaceous. Made up of clay and calcium carbonate.

arid. Dry; having no moisture, barren, applied to areas having less than 15 inches rainfall per year.

arrastre [sometimes, arrastra] (Mexican). A crude machine for pulverizing rock frequently used by early Spanish miners; consists of a circular pit lined with rocks in which ore is broken by big rocks attached to poles fastened to a central axis; motion was supplied by burros, mules or oxen.

arroyo (Mexican). Small or large stream; deep, dry gulley; a wash in desert areas.

artesian well. Originally applied to a well which produced water at the surface from underground pressures; now often used to describe any deep well even though a pump is necessary.

artifacts. Man made implements or objects used by men at least 100 years ago and ranging in age to thousands of years.

assay. To test suspected ores for mineral content, or money value, by chemical, heat or other means.

assessment work. Annual work required to be done on an unpatented mining claim to retain legal possession.

asteriated. Used to describe a transparent or transluscent gemstone which shows a star; i.e., star ruby, star sapphire, star quartz.

asymmetrical. Having no proper proportions; unsymmetrical.

asymmetrical fold. A fold in which one limb dips more pronounced than the other.

atmosphere. Air; the gaseous elements surrounding the earth.

atomic weight. Weight of an element as compared to a standard; for this comparison, hydrogen is sometimes assigned the weight of one; however, oxygen is more frequently used at a weight of 16.

attitude. Used to describe the position of stratum to a horizontal plane; attitude is described by its dip, strike, plunge, etc.

attrition. Rubbing together, wearing down, abrasion, erosion.

auriferous. Containing gold.

avoirdupois. Common trade system of weights used in the United States and England. Sixteen ounces equals 7,000 grains. An avoirdupois pounds equals 14.583 troy ounces or 453.6 grams.

b

bajada (Mexican). A broad alluvial fan extending from a cliff or steep mountain range.

ball mill. A large, rotating cylinder in which balls, usually iron, are used to pulverize non-metallic ore.

bar. A bank of sand gravel, or other material which rises from the bed of a stream or river; a placer deposit usually found in the slack part of a stream.

barchan. A crescent-shaped sand dune with the convex side facing the wind.

barium. One of the rare, heavy metals.

barranca. (Mexican). Common term used to describe any type of deep canyon.

barren. Used to describe veins or terrain that have no minerals of any commercial value.

barrier beach. A low beach which is separated from the mainland by a body of water.

basalt. A general term used to describe almost any fine-grained, dark-colored igneous rock.

base level. Generally the lowest level to which any terrain can be reduced by running water; sometimes the level of any body of water into which a stream or river runs.

basic rock. A term now much in disfavor with geologists and rapidly becoming obsolete; it is generally used to describe an igneous rock with a low percentage of silica but can mean several things.

batea (Mexican). A conical shaped gold pan used in South America.

batholith. A huge mass of intrusive igneous rock; technically it should be at least 40 miles in diameter and have no known depth.

bauxite. The principal ore of aluminum; a rock containing aluminum hydroxides.

bedrock. Solid, unweathered rock covered by unconsolidated material such as sand in a stream bottom.

benches. In mining, used to describe deposits shaped like steps or terraces; i.e., bench placers are terraces from 50 to 300 feet above the present stream level.

biotite. A magnesium-iron mica often found in igneous rocks; it usually forms dark, almost black, crystals.

birefrigence. Double refraction; property of minerals outside the isometric system that results in separating light into two rays.

bituminous coal. Soft coal.

black light. A light produced by ultraviolet radiation used by prospectors to detect fluorescent minerals.

black sand. Heavy grains of magnetite, chromite, ilmenite, etc., which are often concentrated in panning.

blind deposit (Also, blind vein). A mineral deposit or vein which is hidden beneath overburden.

block diagram. A three-dimensional illustration showing the surface features and the geological features under the ground.

block mountain. Mountain or range formed by large, uplifted sections of earth and marked on one side by a fault.

blown sand. Sand which has been formed, transported or deposited by wind; eolian sand.

boleo (Mexican). Float mineral found some distance from its original source.

borehole. A hole drilled with an auger, drill or diamond bit for prospecting or blasting purposes.

boulder. Any rock with a diameter of more than ten inches.

brackish. Used to describe undrinkable water; salty water.

breccia. Rock characterized by angular or sharp fragments as opposed to waterworn conglomerates.

brittleness. The tendency of a mineral to break easily.

broken. Veins which have been discolated by faulting.

bull-pup. A mining claim which is worthless.

bullion. Gold or silver, either alone or combined in metal form.

bunchy. Said of an ore that has small scattered masses of ore.

butte. A solitary hill or mountain which has extremely steep sides.

button. The small mass of metal remaining after fusion or assay.

C

calcareous. Material made up of or containing, calcium carbonate; limy.

calcite. crystallizing calcium carbonate; more commonly, limestone or chalk.

caldera. A large crater formed by the explosion of a volcano.

calyx. A drilling machine which cuts a core that can be preserved as a sample.

Cambrian. The oldest period of the Paleozoic Era.

carbonaceous. Material which contains carbon.

canyon. A valley or gorge usually with precipitous sides.

carat. A unit of weight used for weighing diamonds; also employed to indicate the fineness of gold; i.e., 24 carat equals pure gold.

carbonate. Rocks with a high percentage of carbon dioxide.

carborundum. A trade name for artificial abrasive or grinding material; used primarily for sharpening tools or smoothing metal.

Carboniferous. The fifth geologic period of the Paleozoic Era in Europe; American geologists use the Mississippian and Pennsylvanian Periods to describe the same time.

carbonization. In geology the process whereby organic material is fossilized to carbon or coal.

caving system. A method of mining where

the ore is broken by deliberately letting it cave in.

cement. Material which binds together pieces of sedimentary rocks; sometimes used to describe such rock.

Cenozoic. Geological time immediately following the Mesozoic Era and extending to the present; latest geological era.

center cut. The bore holes put in a face to include a wedge-shaped piece of rock which are fired first in blasting ore.

cerro. (Mexican). A hill or mountain.

chalk. Soft, white limestone.

chemical weathering. The process of weathering which results in a change in chemical composition.

chert. A very hard, glassy mineral; flint-like variety of silica.

Chilean mill. A grinding mill composed of vertical rollers running in a circular enclosure with an iron or stone base which is used for fine grinding ore.

chimney. An ore shoot running vertically.

chlorides. A common term used to describe ores containing chloride of silver.

chlorite. In geology, a term used to describe a greenish mineral which is extremely soft.

cinder cone. A steep, volcanic cone composed primarily of ash and cinders.

clinometer. An instrument used to measure angles, especially dips of veins.

cirque. A deep, bowl shaped depression in a mountain caused by glacial action.

claim. In mining, that land which may be legally mined by the person who has the right to it.

clastic rock. Rock that has been moved from its original location.

cleavage. In geology, the properties of certain minerals to be split parallel to its crystal planes.

coastal plain. A large level section of terrain near a big body of water; usually the portion of the sea floor which has recently been elevated.

col. Saddle-like gap across a ridge where streams flow in two directions or between two mountain peaks.

collar. In a mine shaft, the timbering around the entrance or top.

color. In mining, the shade or tint of the earth or rock that indicates minerals; in gold panning, any particle of gold in the pan.

columnar section. An illustration which shows geological formations, their thickness and order of accumulation.

columnar structure. A mineral containing parallel rod-like crystals.

comminute. In mining, to reduce ore to powder by grinding.

compaction. In geology, used to describe the method or process by which materials are cemented into sedimentary rocks.

complex. In geology, literally to contain many minerals; an ore is complex when it contains many different metals.

composite cone. The cone of a volcano which was formed by alternate layers of cinders and lava.

compound vein. A vein which contains several minerals; also a vein which is crosscut by several fissures, usually diagonally.

concretion. An aggregate of mineral matter which forms around a central core.

concentrate. Separating valuable minerals from their matrix or carrying agent.

confluence. The place where streams join.

conglomerate. A cemented clastic rock which is formed by rocks of pebble size.

connate water. Water which was trapped in rocks during their formation and has never been surface water.

consolidation. In geology, the procedure in which sediment is changed into rock.

contact. In mining, used to describe the point where two bodies of rocks or minerals join.

continental shelf. That shallow portion of an ocean or sea which surrounds a continent.

contour interval. On geological maps, the difference in elevation between two consecutive contour lines.

contour line. A line drawn on a geological map to mark points of the same altitude above sea level.

costeanning. Finding, or proving a lode deposit by trenching across it at right angles; also to remove overburden by force of water for the same purpose.

core. In geology, the innermost part of the earth; in mining, a cylindrical portion of earth obtained for a sample by drilling.

correlation. Generally, the process where two geological formations are determined to belong to the same geological age.

country rock. Commonly the predominant rock of any region; in mining, usually means the rock which ore penetrates.

coyote hole. A small tunnel drilled away from the main tunnel, at the ends are two crosscuts which are filled with dynamite with the resulting explosion blowing out

167

large masses of rock; same as gopher hole.

crevasse. A deep chasm in a glacier, snow, ice or a deep split in the earth after an earthquake.

crevice. In mining, a shallow split in the bedrock of a stream where gold is often concentrated; in Colorado mining law, a mineral bearing vein.

crib. A method of timbering mines in which logs or timbers are laid one upon another.

crosscut. Commonly, a tunnel which crosses another tunnel; a tunnel which crosses, instead of follows, a vein.

cross-cutting relationships, Law of. A geological law which states that a rock is younger than any rock across which it cuts.

crushed vein. A vein made up of crushed material which was generally caused by folding, faulting or shearing.

cryptocrystalline. Mineral composed of such fine crystals that they cannot be seen without the aid of a microscope.

crystal. A common mineral form which is characterized by several flat surfaces in regular form. Crystals have definite external symmetry and internal structure.

crystalline. Used to describe a mineral which has a crystal like internal structure but not necessarily such an external structure; opposite of amorphous.

cube. In geology refers to a six-sided crystal form each face of which is perpendicular to the crystal's internal axis; differs from mathematical cube in that it need not have equal sides.

cupola. In mining, a small blast furnace used to refine ores.

cusec. A unit of measurement used to indicate the flow of one cubic foot per second of water or air.

cyaniding. In mining, a process where gold or silver ores are combined with sodium or potassium cyanide to obtain metals.

d

dacite. An igneous rock made up of feldspar and quartz.

deadman. A wooden block used on rails to guard against runaway cars; also a log or other buried item serving as an anchor for block and tackle.

debris. Tailings; fragments from a blast; fragments at the base of a cliff.

decomposition. The chemical breakdown of rocks.

deflation. The process of removing sand or loose material by the wind leaving the rock surface bare.

deformation of rocks. The distortion of rocks by pressure, usually from folding or faulting.

delta. A level area formed where a stream enters a body of standing water, usually triangular in shape with the point headed upstream.

dendrite. A branch or treelike figure produced on or in a rock or mineral; usually formed by crystallization of oxide of manganese or other foreign material; i.e., an example would be moss agate.

dendritic. In geology, branching like a tree; some forms of crystallized gold are described as dendritic.

deposit. In mining, means any natural occurrence of commercial quantities of minerals; an occurrence of metal concentrated in a stream bed.

deposition. In geology, the process of natural accumulation of a mineral or metal by action of wind, water, or other causes.

desiccation. The loss of water or drying out of any substance.

detrital. Used to describe minerals occuring in sedimentary rocks that was derived from earlier sedimentary, igneous or metamorphic rocks.

development. In mining, the process of working an ore deposit.

Devonian. Geological time of the Paleozoic era; period which follows the Silurian and precedes the Mississippian.

diamond. Nature's hardest known substance, crystalline carbon.

diastem. A minor break within a geological formation usually represented by a bed or beds.

diastrophism. The process in the earth's crusts which creates major earth formations such as ocean basins, plateau's, mountains and etc.

diatomite; diatomaceous earth. A slate which is made up of fossils (diatoms) and pulverized; it is used as a filter and in other commercial applications.

dike. In geology, a tabular igneous rock body which when molten, cuts across the original rock strata.

dip. In mining, the angle of inclination which the vein makes with a horizontal plane.

discovery. The original finding of a mineral deposit.

disintegration. Breakdown of rocks by weathering or chemical means.

dissected. In geology, areas cut into hills and valleys by erosion.

disturbance. The bending or faulting of a rock or stratum from its original position; also a geological event which spans two periods.

divide. A ridge or high area which separates adjoining drainage areas.

dobie. Slang for adobe; clayey; muddy.

dome. A symmetrical hill or mountain which resembles a dome in a building; in geology, the opposite of a basin.

double refraction. Birefrigence; the capability of a mineral to split a ray of light into two separate rays.

downcast. A shaft through which air is provided to lower levels of a mine.

drift. Material deposited by a glacier; in mining, a horizontal passageway which follows a vein as opposed to a crosscut which intersects the vein.

drumlin. A rounded, streamlined hill composed of glacial drift with its longitudinal axis in the direction of the glacier's travel.

drusy. Cavities in mineralized veins or ore usually covered with very small crystals.

dune. A sand hill or ridge formed by the wind.

dunite. A rock of the serpentine family.

e

earthquake. A trembling, shaking, undulating or sudden shock of the earth's crust; earthquakes are sometimes accompanied by fissures in the earth's surface or giant waves when they occur beneath the ocean.

efflorescent. An incrustation on rocks that often resembles lichens

electrum. A very rare natural alloy of about 75 percent gold and 25 percent silver; obsolete name for amber.

element. A fundamental substance of the earth; a substance which cannot be broken down into other substances.

eluvium. Usually sand accumulated by wind; similar material collected by water is called alluvium.

end moraine. An accumulation resembling a ridge at the margins of a valley glacier or ice sheet.

enhydrous. Containing water; rocks or minerals which contain drops of fluid.

Eocene. A division of geologic time of the Cenozoic Era; follows the Paleocene and precedes the Oligocene.

epicenter. In geology, the point directly above the focus of an earthquake.

epoch. In geology, a division of time; subdivision of a period.

era. In geology, a division of time; includes one or more periods.

erosion. Wearing away, removal or breakdown of any earth substance by natural methods; i.e., wind, water, waves or ice.

erratic. Pebbles, rocks or boulders that were transported from their place of origin by a glacier and left on alien bedrock when the ice melted.

eskers. A long ridge of sand or gravel deposited by a stream running through or under a glacier.

estuary. The mouth of a river where it meets the ocean tide; a drowned river mouth caused by the rising of the sea floor; a large pool caused by the effects of the tide.

evaporation. The process by which a liquid is changed into a vapor and carried away.

evaporite. A sediment that results from the evaporation of a liquid; i.e., salt from salt water.

exfoliation. The process which results in a splitting off of scales, slabs or flakes from rocks during weathering.

exposure. In geology, refers to unobscured outcrops of rocks or minerals at the surface of the earth.

extrusive rock. Igneous rock that has been forced out of the earth as opposed to the same rock forced into other, older rocks.

face. Any outer flat surface of a crystal.

facet. A small face; small, flat, regular surfaces; flat surfaces on a gemstone.

fault. A fracture or fracture zone along which there has been a movement of the two sides parallel to the fracture; a fault can move from a few inches to several miles.

fault breccia. Breccia found along the plane of a fault.

fault plane. The point where the fault actually moved if it is in a straight line; see fault surface.

fault scarp. A small cliff formed at the surface by one section of the fault being uplifted.

fault surface. The point where the fault actually moved if it is in a curved line; see fault plane.

feldspar. A group of closely related, rock

forming, silicate minerals; includes ortho-clase, labroadorite, plagioclase and others.

fiber. In geology, a fine, thin, thread-like structure found in many minerals.

fineness. Term used to describe the purity of gold; i.e., 948 fine gold contains 948 parts gold and 52 parts of other material.

fissility. The tendency of certain rocks to split easily along closely parallel lines; i.e., slate, mica, schist, etc.

fissure. A large crack or break in rocks or veins.

flint. A variety of quartz composed of silica, it has excellent splitting tendencies and leaves sharp edges when broken.

float. Loose or scattered pieces of ore broken off from the original lode and transported some distance to an alien terrain; placer gold is not referred to as float.

flotation. A process whereby finely crushed ore is separated in a froth caused by chemicals which allow some minerals to float and others to sink.

flood plain. A low, usually flat terrain near a river which is dry at normal levels and covered with water when the river is in flood stage.

flouring. Used to describe the breakdown of mercury into extremely small globules, at which point it will not rejoin and is worthless for amalgamation.

flume. In mining, an inclined channel or trough used to transport water to the place of mining.

fluorescence. The property possessed by some minerals to emit a glow or light during exposure to ultraviolet light.

fluvial deposit. A sedimentary deposit laid down by a stream or river.

flux. In mining, a chemical or mineral added to an ore to enable reduction by heat.

focus. In geology, the true center of an earthquake.

fold. A bend or wave in rocks or strata caused by pressure; the structure of rocks or strata that has been bent into a dome, basin, terrace or roll.

fold mountains. A mountain or mountains that were formed from folding of the earth's crust.

foliated. In geology, a mineral consisting of thin leaves such as mica.

foliation. The banding or lamination of metamorphic rocks as opposed to the term stratification of sedimentary rocks.

fool's gold. A form of iron pyrite which is often mistaken for native gold.

footwall. The wall, floor or rock beneath a vein.

formation. A term used to describe or map large geological features which have the same general distinctive makeup.

fossil. The remains, traces or impressions of plants, animals or other organisms that have been preserved in the earth's crust.

fracture. In geology, the texture and appearance of a freshly broken surface of a mineral; the fracture is one method of determining the type of mineral being examined.

fragmental rock. Rock that has been moved from its place of origin; clastic rock; sometimes used to describe different rocks cemented together.

free gold. Gold which is not combined with any other mineral or metal; placer gold.

free milling. Used to describe ores that contain native gold or silver which can be reduced by crushing and amalgamation without chemical or roasting process.

fumarole. A vent or hole in or near a volcano from which hot gasses issue.

g

gabbro. A fine-grained, dark-colored, igneous rock consisting primarily of plagioclase, feldspar and pyroxene.

galena. Commonest of lead ores; occurs as gray, cubic crystals.

gangue. The nonvaluable minerals or rock associated with a commercial ore.

garnet. A group of silicate minerals characteristic of some metamorphic rocks; often found as an isometric crystal with many shades of red.

gash vein. A vein characterized by being wide at the top and narrow at the bottom without any great depth.

geanticline. A major upward thrust of the earth's crust from which the sediments have been eroded.

geochronology. A system of dating geological events in relationship to the earth's history.

geobotany. The process of prospecting by making a visual inspection of the plants growing in any given area.

geode. A rock with a hollow or filled interior; the cavity is commonly filled with quartz or calcite or lined with crystals of the same materials.

geomorphology. The science dealing with the earth's surface and sub-surface geological features.

170

geophysics. The science dealing with the earth's structure, composition and development including the atmosphere and hydrosphere.

geosyncline. A great, generally linear, downward fold of the earth's crust.

geyser. Literally a miniature volcano which emits columns of steam and water instead of lava and ashes; found only where volcanic action is on the decline.

geyserite. A hydrated silceous deposit formed around the openings of some geysers; a variety of opal.

glacier. A stream or sheet of ice formed by recrystallized snow; glaciers move downhill from the center of accumulation.

glance. A term used to describe minerals with a lusterous appearance; i.e., lead glance for galena.

glass. In geology, used to describe minerals which are non-crystalline and have cooled rapidly; i.e., volcanic glass is actually obsidian.

globule. A small mass, usually semi-shperical in shape, used to describe the residue of an assay test.

gneiss. A coarse-grained metamorphic rock in which the light materials are separated from the dark by mineral bands.

gophering. Prospecting without any apparent objective, such as looking for float on the surface or veins below the surface with no visible outcrop; also, to mine the richest ore in a haphazard way.

gopher hole. Same as coyote hole; to blast off large portions of the face of an ore body.

gossan. An iron deposit filling the upper portion of veins or covering deposits of pyrite.

gouge. A layer of soft material in or along the wall of a vein which can be gouged out with a miner's pick.

graben. A long, narrow depression that is bounded on two sides by faults; a trough fault.

gradient. The rate of ascent or descent of a streamslope or roadway.

granite. A coarse grained igneous rock made up of quartz, feldspar and mica or other colored minerals.

grit. Coarse grained sand or sandstone suitable for use as a grindstone.

grizzly. A heavy iron grating used to screen large pieces of material out of ore being sluiced.

ground water. Water at and below the water table level; all underground water.

h

habit. The characteristic crystal shape of a mineral.

hackly. Applied to minerals which fracture with sharp, jagged edges.

halite. Impure common salt found in cubic crystals; rock salt.

hanging valley. A tributary valley which has a greater elevation than the valley into which it runs.

hanging wall. The country rock above a vein that is being mined.

hardness. In geology, a testing device to determine the ability of a rock or mineral to be scratched.

hardness scale. The standard by which minerals are tested to identify them; the most common scale in use today is Moh's scale which ranges from talc, one, to diamond, 10.

hardpan. In placer mining, used to describe cemented or firm clayey material that is hard to break up for sluicing.

headframe. The framework erected over vertical shafts for the purpose of hoisting ore out; gallows frame.

heads. The valuable portion of the ore that is being mined or milled; the opposite term for the waste is tails.

hematite. One of the most common ores of iron, contains about 70 percent iron.

high-grade. A very rich ore; also to steal rich ore from a mine during the process of mining.

hitch. Holes cut in the sides of a mine tunnel to hold timbers.

hogback. A ridge or line of hills extending in a line and characterized by steep sides.

hornblende. A mineral belonging to the amphibole group that forms with many other substances and the name is attached to several types of ore.

horse. In mining, refers to a large mass of country rock contained within a paying ore.

hot springs. A natural flow of hot water as opposed to a geyser.

hydrated. A mineral which contains water in its structure.

i

Ice Age. The Glacial Epoch; a geological time when many glaciers covered the earth.

ice cap. A continuous covering of ice, usually many feet thick.

ice sheet. A covering of ice that is generally confined and not mature enough to be called a glacier although it may be moving.

igneous rock. One of the major classifications of rocks; rocks that were formed by solidification from a molten state.

impregnated. Used to describe metals or minerals scattered through a vein or country rock.

inclusion. A rock or mineral fragment enclosed by another rock or mineral.

intermittent stream. A stream which flows only at certain times of the year; it may be wet or dry most of the time depending upon the weather.

intrusion. In geology, a mass of igneous rock which was forced into other, older, rocks while in a molten condition.

iridescence. In geology, the property of some minerals to show colors inside themselves; i.e., fire opal. ●

j

joint. A division or seam in a rock along which there has been no movement by the opposite sides parallel to the break.

Jolly balance. A delicately balanced spring scale used mainly for determining specific gravity by weighing in water and air.

Jurassic. A geologic time; the middle Period of the Mesozoic Era.

jackhammer. A hand held, compressed air rock drill that is automatically rotated.

jade. A hard, tough mineral generally green in color; it is highly prized by gem cutters.

jet. Dense, black lignite often used for jewelry.

k

kame. Any small hill or ridge which has been formed by a glacier.

kaolin. A white or nearly white clay containing feldspar, it is used in the making of ceramics and porcelain.

Karst topography. Named for a section east of the Adriatic Sea; characterized by deep sinkholes, caverns eroded out of limestone, streamless valleys, and streams that end abruptly and drain underground.

kettle. An inverted conical basin formed when buried masses of ice melt.

kindly. A miner's term to describe rock or veins that are likely to, or appear to, carry commercial ore.

knob. In geology, used to describe a round, isolated hill.

l

laccolith. A large mass of igneous intrusive rock that has pushed up the overlying rocks.

lacustrine deposit. A mineral deposit which has been formed by a stream leading into a swamp, lake or lagoon.

landslide. The movement of earth, rocks and debris which have been loosened by water, or weathering, down a mountain or hill; the movement may be instantaneous as down a steep mountain, or slow as in the case of large masses of mud moving down a gentle slope.

lava. Term used to describe either the molten or solidified material erupted from volcanos.

leaching. Extracting soluble metals from ores by dissolving them in a suitable chemical such as water, sulphuric acid, etc.

leaf gold. Very thin specimens of native gold usually found between layers of schist.

ledge. Commonly, a very rich outcrop of ore; a narrow vein of gold ore found on the surface.

level. Horizontal passages in a mine from which crosscuts and drifts are cut to obtain ore; levels are in regular spaced intervals, usually 100 feet.

lignite. Soft brown coal which is only a little harder than peat.

limestone. A sedimentary rock containing at least 50 percent of calcite or dolomite.

limonite. An important ore of iron.

lithification. The process by which sediment is changed into rock.

lithology. The science of describing a rock based on a visual inspection with the naked eye.

load. In geology, used to describe the amount of material that can be transported by a natural eroding agent at any given time.

m

magma. A term used to describe the molten material beneath the crust of the earth from which all rocks were originally formed.

malleable. The quality possessed by certain metals to be pounded and flattened without breaking.

mantle. Generally, the thick layer of material that extends from the earth's crust to its core; also loose material or soil on the surface of many deposits; overburden.

marble. In geology, a metamorphic rock made up essentially of calcite and dolomite;

generally, alabaster, onyx, travertine, serpentine and some granites.

massive. In mineralogy, a mineral without apparent crystallization.

matrix. The rock or material in which another rock or mineral is enclosed; gangue; also the rock in which the base of a crystal is embedded.

meanders. In geology, used to describe a series of wandering curves in the course of a stream or river.

mesh. In mining, the number of openings per inch in a screen through which material is sifted.

Mesozoic Era. A geological time consisting of the Triassic, Jurassic, and Cretaceous periods.

metamorphic rock. Any rock which has been changed by great heat or tremendous pressures causing new minerals or different structure to occur; i.e., clay into slate.

meterology. The science which deals with the atmosphere and changes in it.

metalliferous. Specifically means yielding or producing metals but is generally used to indicate any ore body containing a valuable mineral.

mica. A group of silicate minerals having similar properties and nearly perfect cleavage.

mineral. An inorganic substance possessing definite chemical and physical properties that occurs in nature; minerals such as fossils may have been organic in origin.

mineralogy. The science which deals with the occurrence, composition, etc., of minerals.

mineral right. A right, sometimes exclusive of ownership of the surface land, ·to enter and remove the minerals beneath it.

miner's inch. A term used to measure water for the purpose of mining, it varies from area to area and usually indicated between 2,000 and 2,600 cubic feet per 24 hours.

monitor. In hydraulic placer mining, a high pressure nozzle used to direct water on ore.

monocline. Said of strata which dips in only one direction.

monoclinic. A crystal system in which two of the faces are at right angles and the third inclined.

monolith. A single stone, usually very large and tall, shaped like a pillar and standing alone.

monument. A permanent object, usually a pile of rocks, which indicates the ownership or boundaries of a mining claim.

moraine. An accumulation of earth, sand and rocks carried by and deposited by a glacier.

morphology. In geology, the study of the shape and form of land.

mother lode. The major lode or vein passing through a mining district or area.

n

native. In mining, a mineral that occurs in ore as pure metal; i.e., copper, gold, silver or platinum.

neutral. In geology, neither acid nor alkali.

neve. Partially compacted granular snow which covers the top of a glacier; an ingredient of a glacier.

nugget. A lump of native copper, gold, silver, or platinum; only waterworn pieces of ore are nuggets, native lode metal is not so named; to be a nugget the piece should be at least as large as a grain of wheat.

o

Oligocene. A geologic time; part of the Cenozoic Era.

olivine. A magnesium iron silicate of which the name is applied to many rocks containing a percentage of the mineral; a green colored variety is highly prized as a gem stone.

oolite. Small, spherical minerals which may or may not be cemented together to form rocks.

opal. An amorphous hydrous silica which is common; gem varieties such as fire opal, have the ability to show colors in their interior and are very valuable.

Ordovician. A geologic time; the second period of the Paleozoic Era.

ore. A natural mineral compound of which at least one of the elements is a metal.

outcrop. The part of a mineral formation that appears at, or very near, the surface of the earth.

outwash fan. A broad flat area formed by deposits created by streams of a melting glacier; also called an outwash plain.

overburden. In mining, usually used to indicate worthless rocks or material covering a valuable vein; in placering, the non-gold bearing sand above bedrock where gold is concentrated.

p

pay ore. Those parts of a vein with a commercial value; also pay streak.

peat. Partially decomposed plant material; the first stage in the process of making coal.

pegmatite. A coarse-grained, igneous rock which often contains commercial minerals, occasionally these are rare earths.

peneplain. A former plateau which has been lowered by erosion; a relatively large, flat area that once was much higher.

Pennsylvanian. A geologic time; the sixth oldest Period of the Paleozoic Era.

penstock. In mining, a sluice or gate restraining water.

permeable. In geology, refers to a rock that due to its structure permits liquid or gasses to pass through it.

Permian. A geologic time; the last Period of the Paleozoic Era.

peter out. In mining, used to describe a mine or vein which gradually becomes less valuable due to lower grade ore; also to pinch out.

pig. In refining, an ingot or crude casting of any metal to make it more convenient to handle and transport.

pinched. Used to describe a vein that narrows with depth.

pipe. An elongated, usually vertical, relatively small body of ore; an ore shoot.

pitch. In geology, the angle between the axis of a fold and the horizontal plane.

Playa (Mexican). An area in the desert where water is accumulated after rain and then evaporates; a shore, strand or beach of a river; sometimes, a dried up desert lake.

pocket. A small area of very rich ore either isolated or in a vein with lower quality ore; also called a kidney.

porosity. In geology, a descriptive term describing the percentage of any rock which contains open space.

porphyry. Now used to describe all rocks characterized by conspicuous phenocrysts in a glassy or fine-grained ground mass.

pothole. Literally, a small, rounded hole; can be in rock, ground, stream beds, or other terrain.

Precambrian. A geologic time; the time before the Cambrian.

precipitate. In mining, to obtain mineral or metal values from ore by chemical methods.

Proterozoic. A geologic time; first Era of the Precambrian.

q

quartzite. A quartz rock that has metamorphised from sandstone; used by miners to describe any drill-resistant sandstone.

Quaternary. A geologic ·time; the first Period of the Cenozoic Era.

qualitative. In mining, used to describe how many and what metals are in any ore; results of an assay test.

quartering. A process of reducing a sample by fourths for the purpose of obtaining a representative assay of an ore.

quicksilver. Common name for mercury.

r

rake. In mining, the angle of inclination of a vein.

reagent. A chemical or substance used to produce a desired chemical action in treating ore.

recumbent fold. A fold in which the axis of folding is essentially horizontal.

red beds. Used to describe red sedimentary rocks, usually sandstone or shale.

reef. An Australian term meaning vein or lode.

refining. The process of reducing crude ore to its mineral or metal; such as gold sulfide to gold.

refractory. In mining, a complex ore difficult to refine; i.e., an ore containing copper, lead, silver and zinc is refractory.

roast. In mining, to heat ore to the point just short of fusion to effect oxidation so that the ore can be smelted·

rock. Any naturally formed aggregate or mass of mineral forming an essential part of the earth's crust; stone.

rock salt. Common salt; halite.

rockslide. A rapid downward movement of recently detached rocks.

runoff. Rain water that cannot be absorbed into the earth and runs off.

rusty gold. Free gold that is covered by an oxide of iron and will not amalgamate.

s

schist. A crystalline rock which can be readily split due to a parallel structure; also a rock which occurs in thin layers.

sediment. Solid material, either mineral or organic, which has been, or is being, transported and deposited on or under the earth's surface.

174

sedimentary rock. A rock formed from sediment by pressure, cementing or other means.

shaft. In mining, a hole deeper than it is round.; a hole dug for exploration or exploitation of an ore body; a shaft is vertical, tunnels and adits are horizontal.

shoot of ore. Ordinarily a small body of ore that moves out vertically from a horizontal body of ore; a shoot of ore is not very often extensive in size.

single jack. A light, hand-held hammer used by miners in drilling rock.

sinkhole. A depression in the earth usually caused by collapse of an underground area worn away by water; large areas are called sinks.

slate. A fine grained, metamorphic rock formed from shale.

slickensides. The polished stirations and groves on the sides of faults.

slip. Miner's slang for fault.

slope. The incline of a tunnel, vein or entry to a mine.

spoon. A small diameter rod with a spoon like attatchment on the end used for cleaning bore holes or cleaning rich ore from between rocks.

stalactite. An icicle shaped column descending from the roof of a cavern formed by solutions dripping from the roof.

stalagmite. An icicle shaped column ascending vertically from the floor of a cavern formed by minerals remaining after evaporation of water dripping from above.

stope. An often misunderstood mining term; it means to mine an ore body in a series of steps; generally the lower level is mined to a convenient distance and then the next level and so on until the ore body is removed but there are many methods.

strike. In geology, the course or bearing of a line formed by the intersection of a vein or bed with a horizontal plane; strike is perpendicular to the dip.

stringer. A small vein or ore body.

stripping. The process of removing overburden.

sulphate. A salt of sulphuric acid most of which are soluble in water.

sulphide. A combination of sulphur and another mineral; such as silver sulphide.

superposition. A theory which states that if more than one stress operates on a structure each stress may be superimposed to give results of the strain; more commonly a method used to describe how rocks are placed one above the other in layers.

suspension. In mining, the manner in which a moving stream suspends a heavier than water object or metal while it is being transported.

t

tabular. In geology, a mineral having a large flat surface.

tailings. Residue of ore which contains no mineral value; now legally known as debris.

tails. That part of ore which contains minerals that are uneconomical to refine.

talus. A heap of rock debris which collects at the bottom of a cliff.

tarn. A small lake formed in a crique or glacial valley.

tarnish. In geology, a thin layer of color that forms on a mineral.

tectonic. Describes rock and surface formation created by changes in the earth's crust.

tectonic earthquake. Earthquake caused by faulting in the earth's crust.

tenacity. Mineral identification term; describes the resistance of minerals to breakage.

terminal moraine. Furthest advance of a glacier; end moraine.

termination. In geology, in describing a crystal, the end that is completely enclosed by crystal faces.

Tertiary. A geologic time; the oldest Period of the Cenozoic Era.

texture. Mineral identification term; describes appearance of a rock by size, shape and arrangement of materials.

theodolite. A surveying instrument consisting of a tripod, compass, level, scale and telescope.

thrust fault. A reverse fault with a low angle of inclination.

tombolo. A finger or bar of sand which connects an island with the shore or another island.

topographic map. A map showing the physical features including culture and elevation by means of symbols and contour lines.

transportation. The method by which minerals are naturally carried and deposited, such as gold in river sand.

traprock. Used to describe dark colored dike and flow rocks especially basalt and diabase.

travertine. A form of calcium carbonate

sometimes called Mexican onyx, however, it is not a true onyx.

Triassic. A geological time; oldest Period of the Mesozoic Era.

troy weight. A system used for weighing precious metals.

tufa. A chemical sedimentary rock formed around springs and generally composed of calcium carbonate of silica.

tuff. A rock formed of compacted volcanic fragments.

tunnel. Loosely applied to any horizontal shaft but a tunnel must be open to the air at both ends; an adit is open to the air at one end, if it were continued completely through a hill, it would be a tunnel.

twin crystals. Crystals in which one or more parts are reversed in relation to the other parts; twin crystals often appear to be two crystals.

u

unctuous. In geology, used to describe rocks which have a greasy or soapy feeling.

undercurrent. One or more additional sluices which receive the finer mesh material to be sorted.

underhand stoping. The process of mining downward in a series of steps.

underlay. The dip or slope of a vein or strata from the horizontal.

upcast. An opening in a mine through which the air can escape to the surface.

uplift. A large area of the earth's surface which has been thrust upward.

V

valley. A long, V-shaped portion of the terrain between two higher portions of the earth such as mountains or ranges; it commonly has a stream or river near the central portion.

vanner. In mining, a machine for concentrating ore; usually a concentrator table.

variety. In mineralogy, a subdivision of a group of minerals.

varve. Any bed or layer of sediment that is deposited in a period of one year.

vein. A highly localized occurrence of mineralized rock which has significant length and depth but little thickness; vein and lode are commonly used interchangeably but a lode has length, depth and thickness or it is a series of veins interconnected by a common matrix.

vent. In geology, the center of a volcano, or an opening on its sides, where molten lava is erupted.

ventifact. Any stone that has been shaped, smoothed or changed by the abrasive action of wind blown sand.

verdigris. An oxidation of the face of copper commonly called green rust.

vesicular rock. Used to describe rocks, most often of igneous origin, which have small cavities caused by the expansion of gas or steam when the rock was formed.

volcanic ash. The finest particles erupted during the active period of a volcano; volcanic tuff.

volcanic bomb. An isolated piece of magma blown out of a volcano separately which forms a rounded, bomblike shape when it hardens.

volcanic glass. A very hard, non-crystalline rock dark in color with a glassy luster; obsidian is the most common type.

volcano. An opening or vent in the earth's crust which connects with molten magma in the earth's interior and through which the material is forced to the surface, it is commonly a conical shaped mountain made up of the materials erupted; while erupting, a volcano is called active, between eruptions, dormant, and after all activity has ceased, it is extinct.

W

wall. In mining, used to describe the country rock on either side of a vein or lode; above is the hanging wall, below is the footwall.

wash. Used in desert areas to describe a gulley or ravine formed by water but normally dry; washes resemble a dry stream bed; also a mining term used to describe placer mining.

waste. The barren rock removed from a mine; rock which has no commercial value.

water level. The level to which water rises in a mine when it is not removed.

water table. The level under the surface of the earth where water may be obtained by drilling a well; the water table roughly follows the surface of the terrain.

watershed. The elevated and sloping drainage of any high terrain.

windlass. A device consisting of a drum around which a cable or rope is wound to raise and lower materials into a mine; a winch.

winze. A vertical or inclined shaft cut from one level to another in a mine for ventilation or to remove ore.